Essential Guide to Reading Biomedical Papers

Companion website

This book is accompanied by a companion website:

www.wiley.com/go/Langton/biomedicalpapers

The website includes:

Figures and Tables from the book for downloading

Essential Guide to Reading Biomedical Papers

Recognising and Interpreting Best Practice

Editor

Phil Langton

School of Physiology and Pharmacology, University of Bristol, UK

WILEY-BLACKWELL

A John Wiley & Sons, Ltd., Publication

This edition first published 2013, © 2013 by Wiley-Blackwell

Blackwell Publishing was acquired by John Wiley & Sons in February 2007. Blackwell's publishing program has been merged with Wiley's global Scientific, Technical and Medical business to form Wiley-Blackwell.

Registered office: John Wiley & Sons Ltd, The Atrium, Southern Gate, Chichester, West Sussex, PO19 8SQ, UK

Editorial offices: 9600 Garsington Road, Oxford, OX4 2DQ, UK
The Atrium, Southern Gate, Chichester, West Sussex, PO19 8SQ, UK
111 River Street, Hoboken, NJ 07030-5774, USA

For details of our global editorial offices, for customer services and for information about how to apply for permission to reuse the copyright material in this book please see our website at www.wiley.com/wiley-blackwell

Library of Congress Cataloging-in-Publication Data

Essential Guide to Reading Biomedical Papers:
Recognising and Interpreting Best Practice / editor, Phil Langton.
    p. ; cm.
  Includes bibliographical references and index.
  ISBN 978-1-119-95996-0 (cloth) – ISBN 978-1-119-95997-7 (pbk.)
  I. Langton, Phil.
  [DNLM: 1. Biomedical Research–methods. 2. Biological Science Disciplines–education.
3. Research Design. W 20.5]
    610.72′4–dc23   2012028144

A catalogue record for this book is available from the British Library.

Wiley also publishes its books in a variety of electronic formats. Some content that appears in print may not be available in electronic books.

Cover Design: Dan Jubb

Set in 10.5/13pt Times-RomanA by Thomson Digital, Noida, India.
Printed and bound in Singapore by Markono Print Media Pte Ltd.

First Impression   2013

Contents

Companion website

This book is accompanied by a companion website:

www.wiley.com/go/Langton/biomedicalpapers

The website includes:

Figures and Tables from the book for downloading

List of Contributors

Prof Richard Apps, Physiology & Pharmacology, University of Bristol, UK

Dr Nina Balthasar, Physiology & Pharmacology, University of Bristol, UK

Dr Neil Bannister, Physiology & Pharmacology, University of Bristol, UK

Prof David Bates, Physiology & Pharmacology, University of Bristol, UK

Dr Harold A. Coleman, Physiology, Monash University, Australia

Dr Nick Colegrave, School of Biological Sciences, University of Edinburgh, UK

Dr John Crabtree, Physiology & Pharmacology, University of Bristol, UK

Dr Lucy F. Donaldson, Physiology & Pharmacology, University of Bristol, UK

Dr Gordon Drummond, Anaesthesia and Pain Medicine, Royal Infirmary, Edinburgh, UK

Dr Allison Fulford, Centre for Comparative and Clinical Anatomy, University of Bristol, UK

Dr Ingeborg Hers, Physiology & Pharmacology, University of Bristol, UK

Dr Charles Hindmarch, Clinical Sciences, University of Bristol, UK

Dr Mark Jepson, Physiology & Pharmacology, University of Bristol, UK

Prof Michael J. Joyner, Anesthesia Research, Mayo Clinic, Minnesota, USA

Dr Tomoko Kamishima, Cellular and Molecular Physiology, University of Liverpool, UK

Dr Helen Kennedy, Physiology & Pharmacology, University of Bristol, UK

Prof James Ladyman, Philosophy, University of Bristol, UK

Dr Phil Langton, Physiology & Pharmacology, University of Bristol, UK

Dr Thierry Le Bihan, School of Biological Sciences, University of Edinburgh, UK

Dr Fiona D. Mcbryde, Physiology & Pharmacology, University of Bristol, UK

Prof Elek Molnár, Physiology & Pharmacology, University of Bristol, UK

Dr Samatha F. Moore, Physiology & Pharmacology, University of Bristol, UK

Prof Helena C. Parkington, Physiology, Monash University, Australia

Prof Julian FR Paton, Physiology & Pharmacology, University of Bristol, UK

Dr John M. Quayle, Physiology & Pharmacology, University of Liverpool, UK

Dr Scott H. Randell, Cell and Molecular Physiology, University of North Carolina at Chapel Hill, USA

Dr Emma Robinson, Physiology & Pharmacology, University of Bristol, UK

Dr Joshua S. Savage, Physiology & Pharmacology, University of Bristol, UK

Dr Anja Teschemacher, Physiology & Pharmacology, University of Bristol, UK

Dr Paul Verkade, Wolfson Bioimaging Facility, Physiology & Pharmacology and Biochemistry, University of Bristol, UK

Dr Jon Wakerley, Centre for Comparative and Clinical Anatomy, University of Bristol, UK

Dr Erica A. Wehrwein, Physiology, Mayo Clinic, Minnesota, USA

Foreword

Biological research is an experimental science in which the testing of hypotheses through experiments generates knowledge and ultimately, understanding. However, while the construction of a hypothesis is open to the imagination, the experiments that produce the data to be analysed in the light of the hypothesis are only useful if they are reliable, accurate, reproducible, and come from carefully designed and controlled experiments. In the acquisition of data, therefore, there is no room for the imagination and even less for wishful thinking – the vice that reads the data to fit the hypothesis and not the other way around. Therefore, although it is true that a hypothesis guides the gathering of information, the primacy of the data is obvious since without them all hypotheses remain in the "to be done" drawer.

Biological research nowadays can be approached using a variety of techniques, the majority of which have been developed in the last 50–60 years. The power of these new techniques is such that biologists often forget two fundamental things, (1) that the usefulness of the technique does not lie so much in its intrinsic power but in the way it is applied to a biological question and (2) that all techniques have been designed for a specific purpose and therefore have limitations. There is not a single technique that can inform us about the whole of a biological problem and therefore it is advisable, wherever possible, to use more than one complementary technique.

These are the reasons why the present book by Langton and colleagues is so important. It is not merely about techniques; it puts techniques into context – their potential, their limitations and possible pitfalls. This is accompanied by useful primers on the philosophy of science, on experimental design and on statistics. As such, this volume will undoubtedly be of great interest and value, not only to the novice scientist but also to the experienced investigator and mentor.

I highly recommend this book to all practitioners of biological research.
Professor Sir Salvador Moncada
FMedSci, FRS
27th July, 2012

Preface

Imagine you are interested in buying a used car. Are you likely to be entirely trusting of the person selling the vehicle? Will you accept everything they tell you without question and without evidence? I suspect not. I suspect that you will have a *healthy scepticism*. The seller wants your money and it is your responsibility to ensure that you are satisfied with the trade – hence the phrase '*caveat emptor*', which means '*let the buyer beware*'.

I would argue that you should regard journal articles with the same healthy scepticism, but I imagine your reaction to this is either disbelief or a mixture of confusion and panic. If you are sceptical, good! I need to provide evidence for my argument. If you are confused, let me explain.

How is a journal article like a used car? It is easier to consider how the seller equates to the author(s) of a journal article. The seller wishes to get the best price and is unlikely to point out defects and flaws; certainly not with the same enthusiasm as they have for the *plus points*. The authors of a journal article are also selling something – their interpretation of their experiments, including their underlying assumptions. It is hugely important that the authors win the reader's confidence; that they convince *you* that their work is showing true facts (see Primer 1 for an expansion of this idea). Science is big business and careers depend on how widely and how securely the views of the authors are endorsed by the scientific community. In some ways, it is possible to regard journal articles as advertisements for a particular doctrine.

Some of you will have picked up a counter-argument in the shape of the *peer-review process*. Good for you, you are reading with scepticism. Surely the peer-review process means the reader can have confidence that an article in a peer-reviewed journal is accurate and true – at least at the time it was published? Sadly, this would not be a safe assumption. Though it is arguably the best system we have to ensure the quality and validity of what is published, peer-review is very far from perfect and even the most prestigious journals make mistakes.

Science makes use of an approach that has become known as the Scientific Method (see Primers 1 and 2). It was developed into something we can recognize today in a process that can be traced back nearly 400 years to a succession of great thinkers and great philosophers. In 1601, Francis Bacon (1561–1626) wrote:

> *'Read not to contradict and confute; nor to believe and take for granted; nor to find talk and discourse; but to weigh and consider.'*
> From Bacon's essay *of Studies* (published in full in [1]Madden, 2007).

It is too easy to read and simply accept as facts those things that are offered as such. It is highly likely that excellence in your written work, which relies upon your interpretation of the academic literature, will be judged on your ability be demonstrate that you are *'critical of what you read'*.

Unless you have knowledge of the experimental techniques used in a study, however, you will find it difficult to discriminate between studies that are well designed and/or controlled and those that are not, and you will find it hard to be critical. Therefore, in essence, each primer in this book is intended to provide you with the means to be critical about studies described in journal articles.

Aims of each primer

It is important that you are aware of the aims of these primers. Each one is designed to:

- ✓ provide orientation and guidance to readers who have no experience of the technique;

- ✓ suggest reasons/motives for electing to use a specific technique;

- ✓ provide details of a method only where detailed knowledge is required;

- ✓ provide limits on what can 'reasonably' be claimed of data – specificity, selectivity, etc.;

- ✓ describe control experiments that *should* be included in a journal article;

- ✓ explain why particular controls are performed;

- ✓ list other techniques that are commonly used in conjunction;

- ✓ list common misconceptions about a technique or the data produced by a technique;

- ✓ list some caveats about interpretation of data [where appropriate].

---

[1] Madden, P. (Ed., 2007). *Quotidiana* (http://essays.quotidiana.org/bacon/studies/). Accessed 22 Apr 2012.

However, they are *not*:

✗ intended to be encyclopaedic manuals or reviews;

✗ 'how to' guides;

✗ sufficient in themselves as resources (hence the further reading);

✗ likely to be useful to persons experienced with the technique.

I want to end this preface with another quotation from Francis Bacon:

*"If a man will begin with certainties, he shall end in doubts; but if he will be
content to begin with doubts, he shall end in certainties."*
From Bacon's [2]*The Advancement of Learning* (1605) book 1, primer 5, section 8

What I take from this quotation is that it is *not* healthy for scientists to believe too
fiercely in what appears today to be true. We must be prepared to question anything;
there should be no [3]dogma in science, because our current understanding reflects a
continuum beginning with the tentative *'more probable than not'* and moving to
greater and greater probability of being accurate or true – but *never reaching
certainty*. Ultimately, nothing is ever proved.

Phil Langton

[2] Available from: www.lifesmith.com/Berkeley%20Teaching/The%20Oxford%20Dictionary%20Of%
20Quotations.pdf; accessed 22nd April, 2012.
[3] **Dogma** – 'a principle or set of principles laid down by an authority as incontrovertibly true'. Source: Oxford
Dictionaries.

Acknowledgements

This book can trace its origins back to about 10 years, to learning and teaching resources for undergraduate students studying anatomy, neuroscience or physiological science at the University of Bristol and this explains in large part why Bristol academics are involved with the majority of the primers you will find listed. The aim for the book is simple, as it was for the original resource; to communicate some of the practical wisdom that can only come with years of experience and scholarship in laboratory research. It does not pretend to make the reader an expert or to train the reader to use these research techniques but it does provide insight into the assumptions and issues that can lie beneath the surface of seemingly transparent and persuasive research reports. In editing the book, I have been surprised on a daily basis by the scope and significance of the aspects highlighted by my colleagues. Indeed, I have learned so much that I would now view with extreme scepticism the claim that one can read contemporary research reports without, as minimum, the sort of insight that this book aims to provide.

The original resource would not have been possible without the enthusiastic support of my colleagues in the University of Bristol and it is to them that I am most deeply grateful. I wish also to thank the team at Wiley-Blackwell who saw the potential for a book and the independent reviewers who reacted so positively to the preliminary outline and example primers. The first edition required a significant expansion from the original resource and had involved a large number of people from Universities in the United Kingdom, Australia and North America to whom I am enormously and sincerely grateful.

I wish to express my gratitude to the University of Bristol for its encouragement of excellence in learning and teaching as well as excellence in research. For some time the prevailing wind within UK HE has benefitted a focus on research above all else and yet Bristol has striven consistently to promote the interests of its students and the education they receive. This book reflects the ethos of enquiry and excellence that is so typical of the University of Bristol.

I need to acknowledge the support (and patience) of my partner, Rosie, without whom this project would have failed. Finally, I gratefully acknowledge my parents who taught me to respect the potential in dedication and hard work and the value of integrity; attributes that no research scientist should ignore. Lastly to Alice, Polly and Jess who just wanted to see their names in print.

Dr Phil Langton
August, 2012

Introduction

Phil Langton

This introduction explains the structure of most of this publication (Primers 4 to 35). The first three primers escape this structure, as their aim is different. These first three primers cover:

1. The philosophy of science

2. Experimental design

3. Statistics

These are topics of fundamental importance in science. Reading these first will allow maximum benefit from the other primers and indeed from every journal article you read in future. Switch on your scepticism!

As Claude Bernard said in his [1]textbook, *"L'expérimentateur doit douter, fuir les idées fixes et garder toujours sa liberté d'esprit"*, *which means "the investigator should doubt, avoid preconceptions, and always keep an open mind"*

Primers 4 to 35 will have the following structure:

- **Basic 'how-to-do' and 'why-do' section**
  This section will cover what the technique is used for. For some techniques, the answer is far from obvious. If there are facets of the technique that require insight into how the experiments are done, then they will be included in this section. It will be basic – just enough to make sections on *required controls* and *pitfalls in execution or interpretation* intelligible and to help you tackle articles in the reference section.

- **Required controls**
  If your eyes have at tendency to glaze over at the mere mention of *'experimental controls'*, then give yourself a slap and read on. Without well-judged experimental design and comprehensive controls, the results of an experiment are next to useless. There will endless possible interpretations – so much so that the experiment will not advance our understanding. Each

[1] *'Introduction à l'étude de la médecine expérimentale'*, 1865.

primer will list the controls that should be present, and particularly those that require special attention, because these are frequently absent or poorly designed in published articles. If appropriate, the section will indicate what failure of a particular control might indicate. For example, failure of the negative control in a PCR experiment suggests contamination of the RNA by genomic DNA – mRNA implies ongoing transcription, but the presence of genomic DNA says nothing about transcription.

- **Common problems or errors in literature**
  The errors pointed out should be findable by undergraduate students who have little or no practical experience of the technique. This does not mean there are not others – just that you will not be able to recognise them without first-hand experience, i.e. without using the technique to perform experiments.

- **Pitfalls in execution or interpretation**
  If you think that the peer-review process prevents the publication of studies that contain significant errors then, again, give yourself a slap – you're being naïve! Consider this example: a study uses immunocytochemistry (Primer 13) to provide evidence that a protein is expressed in sections of tissue or in isolated/cultured cells; it then goes on to do immunoblotting (Primer 15) to back up the immunocytochemistry (demonstrating that the protein bound by the antibody has the correct apparent mass), *but* uses different primary antibodies for the immunocytochemistry and immunoblots. Yes, it does happen! If the significance of this logical error is lost on you, then you need to read Primers 12 to 15.

- **Complementary and/or adjunct techniques**
  This will list techniques that are often used together.

- **Further reading and resources**
  Some of the articles will be those cited in the text, while others will be suggestions for further reading. You need to keep in mind that these techniques require years of careful scholarship and training to master. The primers have been written by researchers who have developed true expertise with each technique but each primer is very short and of limited scope – hence the term, '[2]*primer*'.

---

[2] **Primer**: a book (or text) that covers the basic elements of a subject. Source: OED.

Section A

Basic Principles

1
Philosophy of Science

James Ladyman

Philosophy, University of Bristol, UK

1.1 What is science?

Dictionary definitions speak of a systematic body of knowledge, and the word '*science*' comes from the Latin word for knowledge. However, not any old collection of facts – even one that is organized – constitutes a science. For example, an alphabetical list of all the words that are used in this book and all the others published on the same day would make no contribution to scientific knowledge. Something else is needed, and there are two obvious supplements to what has been said so far:

- First, the subject matter must be the workings of the physical world. There must be discovery of natural laws and the relations of cause and effect that give rise to the phenomena that we observe.

- Second, the relevant theories must be generated in the right way.

In fact, most philosophers of science and scientists define science in terms of its methods of production; science is knowledge produced by the scientific method. For many people then, asking the question with which we began really amounts to asking, '*What is the scientific method?*'

There are, of course, many methods, and this book is about some of them. The techniques and procedures of the laboratory and experimental trials and the measurement, recording and representation of data, as well as its statistical analysis, form at least as much a part of science as what it tells us about the world as a result. Clearly, the methods of geology and astrophysics differ from those of cell biology or pharmacology.

However, all the sciences we now take for granted have really only reached maturity and separation from each other within the last few hundred years. For example, biochemistry and neuroscience have only become separate disciplines in

Essential Guide to Reading Biomedical Papers: Recognising and Interpreting Best Practice, First Edition.
Edited by Phil Langton.
© 2013 by John Wiley & Sons, Ltd. Published 2013 by John Wiley & Sons, Ltd.

the last century, and whole areas of enquiry were impossible before the invention of electron microscopy and magnetic resonance imaging. Our gigantic science faculties, with their highly specialized disciplines, originated in the ancient and medieval systems of knowledge, and these made very few of the distinctions in subject matter that we now would. for example, many posited connections between the planets and human diseases and other conditions where we find none. Nonetheless, we can find some original truths from many subjects discussed a long time ago. For example, Aristotle recorded that bees pollinate flowers, and the 28-day cycle of the Moon's phases has been known since prehistory.

Modern science is usually regarded as having originated at the turn of the 16th and 17th centuries. At this time, the established ways of predicting the motions of the planets, which placed the Earth at the centre of the solar system, were replaced by the Copernican theory placing the Sun at the centre, which was then modified by Kepler to incorporate elliptical orbits. The latter's laws were precise mathematical statements that fitted very well with the detailed data that had recently been gathered using new optical technology. In the years that followed, telescopes, microscopes, the air pump and clockwork and other mechanical devices were invented and, over the next few generations, knowledge of chemistry, biology, medicine, physics and the rest of what was then called 'natural philosophy' grew enormously.

An amazing thing about all the scientific knowledge that we now take for granted is that the founders of modern science envisaged its production by the collaborative endeavour of people following the scientific method. They argued that there was a common core to all the methods mentioned above, and they advocated the collective use of a single set of principles or rules for investigation, whatever the subject matter. Different people had different ideas about exactly what the method should be, but everyone agreed that testing by experiment is fundamental to science. The task, therefore, is to say what exactly 'testing by experiment' means.

There two general kinds of answer:

- **Positive**, according to which the job of scientists is to gather data from which to infer theories, or at least to find out which theories are supported by it.

- **Negative**, according to which the real task is to try and prove theories false.

The latter may sound strange, but in fact many scientists put more emphasis on it than the former. The reason for that is that there is a very great tendency in human thought to find confirmation of preconceptions and received ideas by being selective in what is taken into account.

The phenomenon known as 'confirmation bias' has been studied extensively in psychology; it is manifested in many ways, including by people selectively remembering or prioritizing information that supports their beliefs. It is very difficult to overcome this tendency, so some people argue that science should always be sceptical and that attempts to prove theories false should be at its heart.

Modern science began with the upturning of many entrenched beliefs about the world, but since then the history of science has repeatedly involved the overturning of cherished doctrines and the acceptance of previously heretical ideas. Examples include the motion of the Earth, the common ancestry of the great apes and human beings, the expansion of the universe and its acceleration, the relativity of space and time, and the utter randomness of radioactive decay. Even the greatest scientific theories, such as Newton's physics and Lavoisier's chemistry, have been subject to substantial correction.

Hence, many scientists follow the philosopher of science Karl Popper in saying that the scientific method consists in the generation of hypotheses, from which are deduced predictions that can, in principle, be falsified by an experiment. When an experiment does not falsify the hypothesis, it may tentatively be employed to make predictions – but the aim should be to seek new kinds of test that may prove it false. A theory that makes specific and precise predictions is more liable to falsification than one that makes only general and vague claims; so, according to Popper, scientists should strive to formulate hypotheses from which very exact statements about experimental outcomes can be derived, and to say in advance what would count as falsification.

Popper emphasized that scientific knowledge is always revisable in the light of new empirical findings, and that science has succeeded in increasing its accuracy, depth and breadth, because even well-established theories are not regarded as immune from correction and revision. Science is not compatible with absolute certainty and the refusal to question.

However, it is also true that in practice, scientists do not immediately abandon core theories when experiments go against them. For example, Newton's law of universal gravitation, the famous inverse-square law, gave beautifully accurate predictions for the paths of the planets in night sky and improved on those of Kepler, as well as generating successful new predictions such as the return of Halley's comet and the flattening of the curvature of the Earth at the poles. However, in the 18th century it was found that the orbit of Uranus was not as predicted, but astronomers did not abandon Newtonian mechanics as a result. Instead, they looked at the other assumptions that they had made in order to calculate the orbit. They had assumed that only the gravitation attraction of the Sun and six other planets needed to be taken into account. If there was another planet, that might explain the anomaly; therefore, Neptune was looked for and found.

Modifying a theory to take account of data that contradicts the original is not, in itself, bad practice. In the case just mentioned, the modification led to a new prediction that could be tested. Science often proceeds like this and, indeed, Pluto was found in the same way. It is now common in astronomy to infer the existence of unobservable objects because of their hypothetical gravitational effect on observable ones.

These examples illustrate an extremely important feature of science, which is that predictions and, hence, tests are never of single hypotheses but always of a collection thereof. To predict the orbit of a planet, one must know all the bodies to whose gravitational attraction it is appreciably subject, and also all of their masses and its mass. If the data do not fit, then logic dictates that there is a problem with at least one of the laws or the other assumptions – although not which one. This is called the Duhem problem (after Pierre Duhem). Scientists face this every day, but they rarely consider that a central theoretical component is false as Popper imagines. To do so would not be sensible, because those core beliefs have been at the centre of a vast number of successful predictions. On the other hand, there will often be many other plausible culprits among the other assumptions involved, and the art and practice of science involves teasing them apart and finding out which to amend.

It is not plausible to argue, as Popper did, that no matter how much a hypothesis has agreed with experiment and survived attempts to show it to be false, there are no positive grounds for belief in it. Since Francis Bacon proposed his new logic of 'induction', many others have sought to develop an account of how evidence can be said to support or confirm a theory. Thus we have two extreme positions:

- *Falsificationism* says science is about showing theories to be false.

- *Inductivism* says science is about showing theories to be true.

It is tempting to seek a happy medium able to incorporate the importance of both, but clearly we cannot do this without some notion of confirmation in science. It is often the case that we look to science to tell us positive facts, such as that a drug is efficacious and safe, or that a particular pathogen is the cause of some medical problem. Bayesian statistics provides measures for how much a given body of evidence supports a given hypothesis. On the other hand, statistical methods are also sometimes used in a falsificationist spirit, as when they are used to calculate the probability of the so-called 'null hypothesis', according to which some potential causal factor has no effect.

The fundamental problem with the scientific method is that it cannot tell us how confident we need to be in a theory before we accept it. Nor, if a research programme is in trouble, can it tell us exactly when to abandon it. For example, in the 19th century, more accurate measurements revealed that the orbit of Mercury did not fit with the predictions of Newtonian gravitation. The trick of positing another planet was tried but, because Mercury is so close to us, any such new planet ought to have been immediately obvious. Thus, it was thought that perhaps it was always the other side of the Sun from us. As it turned out, there is no such planet, and it took Einstein's then new theory of General Relativity to solve the problem.

Similarly, when the evidence begins to come in about the efficacy of a new drug, there is no mathematical formula that can say when we should regard it as '*known*' to be effective. Some scientists may feel sure very early on in the trials, and there may be patients who could benefit from its immediate prescription. However, others will insist that larger studies need to be done before the evidence is compelling. In the end, a committee will set the bar at some level, perhaps demanding that the probability of the null hypothesis for the drug acting on the condition be shown to be less than 0.05 per cent. That is reasonable, but it could also be set at 0.5 per cent or 0.005 per cent, or any other small value and which value is chosen is to some extent arbitrary. Clearly. if the chance of a drug being completely useless is 50 per cent. it should not be prescribed, and if it is .0000000005 per cent then it should be; but where exactly the line should be drawn between these extremes is a matter of choice and judgment.

It is therefore important to be very clear about the limitations of the scientific method, as well as its great power. How much evidence we demand before reaching a conclusion depends in part on whether we are more keen *to have true beliefs* or *to avoid false ones*. If all we care about is having true beliefs, then, for example, above all else we will wish to avoid failing to believe a drug works when it does; if all we care about is not having false beliefs, then, for example, we will wish above all else to avoid believing that a drug works when it does not. The former attitude emphasizes avoiding false negatives and the latter emphasizes avoiding false positives, and in general doing well in respect of one is at the cost of doing badly in respect of the other. Falsificationists emphasize avoiding false positives, so they always think of scientific theories as not yet falsified rather than as confirmed.

The problem is that, both in life and in science, we often need to stick our necks out and commit to the truth of a theory, because if we always wait for one more trial, patients will be denied treatments they need. Part of being a good scientist is developing good judgment about such matters, and it is also necessary to learn where reasonable disagreement is possible, how to identify the crux of such disputes, and how to use the scientific method to refine the evidential basis on which they can be resolved.

Further reading

Bala, A. (2008). *The Dialogue of Civilizations in the Birth of Modern Science*. Palgrave Macmillan.

Ladyman, J. (2002). *Understanding Philosophy of Science*. Routledge.

Popper, K. (1963). *Conjectures and Refutations: The Growth of Scientific Knowledge*. Routledge.

2
Ingredients of Experimental Design

Nick Colegrave
School of Biological Sciences, University of Edinburgh, UK

Well-designed experiments have a central role in life science research. Unfortunately, experimental design is often treated as an afterthought in life science education (all too often appearing as a brief interlude in a statistics course amongst the real business of model fitting and P values!). As a result, it is often viewed with trepidation and is frequently misunderstood. Without an understanding of the basic principles of experimental design, evaluation of the work of others will be typically limited and superficial. Fortunately, these basic principles are not difficult to understand.

Here I focus on five key aspects of experimental design and provide five questions that you should ask when evaluating work carried out by others. For each, I will outline briefly why the answer to the question is important, and how assessment of the value of the study should be modified in light of our answer. Due to the limits of space, I focus on the design of manipulative experiments designed to test causality (i.e. does variable X affect variable Y), considering only in passing studies which lack manipulation. Studies solely designed to estimate parameters, while important, will not be considered at all.

2.1   Is the study experimental?

Suppose that you are interested in whether caffeine intake affects aerobic fitness. You might survey people on their caffeine intake and then measure their performance in an exercise task. A relationship between caffeine intake and performance would then support the hypothesis. This is a correlational study (or observational study). It makes use of naturally occurring variation in one variable and looks at how this relates to variation in the other. Note that 'correlational', in this context,

Essential Guide to Reading Biomedical Papers: Recognising and Interpreting Best Practice, First Edition. Edited by Phil Langton.

does not relate to the method of analysis but to the fact that the variables are not manipulated.

Correlational studies, though important in biology, come with important limitations. The major issue is that their results are often open to alternative interpretations. First is reverse causation; we conclude that caffeine affects aerobic fitness, but perhaps having higher aerobic fitness makes people more likely to drink coffee? Sometimes reverse causation can be ruled out from first principles (e.g. the laws of physics preclude something that happens in the future from affecting something that happened in the past), but this will not always be the case. In this study, it seems implausible but not impossible.

Second, the observed relationship may be due to a confounding variable (i.e. a third variable) that affects both of the measured variables. Perhaps variation in a person's body mass index affects their propensity to drink caffeine and also their aerobic fitness, and this leads to the apparent relationship that we see. Obvious third variables can be measured and controlled for, either in the design or analysis, but there will always be others we cannot measure or have not considered. Put simply, correlation does not mean causation.

Another approach is to carry out an experimental manipulation. We might provide one group with caffeinated drinks and the other with decaffeinated drinks (i.e. we manipulate their caffeine intake), then measure their performance in the trial. Experimental manipulation rules out any possibility that reverse causation could explain the pattern, and since it also decouples the value of the variable of interest from confounding variables (assuming the experiment is carried out properly), the problem of third variables is also removed. Thus, experimental studies generally provide stronger evidence than correlational studies, although in some situations experimental manipulation may not be feasible or ethical.

Take home

A first step in evaluating a study is to decide whether it is correlational or experimental. If it is the former, you should worry about confounding variables and reverse causation and whether the authors account for them. If it is experimental, the next steps are to consider whether the experiment has been appropriately designed.

2.2   Is the study properly replicated?

A central component of any experiment is replication. A single measurement tells us little, but replicate measures, combined with statistical analysis, allow us to decide whether any patterns we see in our data are likely to be real (rather than due to chance).

A key requirement of our replicate measurements is that they must be independent. To appreciate why, consider a situation where they are not. Let's say we are interested in whether a diet supplement affects oestrogen levels in female rats. We take two rats, one fed exclusively on a standard lab diet, the other on the supplemented diet. Even if we measure oestrogen levels in each rat multiple times, we cannot say anything about the effect of the treatment from such a study. The reason is straightforward: rats probably differ in oestrogen levels for many reasons independent of diet. Any inherent differences between the two rats will apply to every measurement we take from these individuals; multiple measurements of the same rat are not independent measures of a treatment applied to that rat – they are what are referred to as 'pseudoreplicates'.

The general problem of treating pseudoreplicates as if they were independent (i.e. the error of pseudoreplication) is that it can lead to us thinking that we have more support for a pattern than we actually do. Essentially, this is because we think our sample size is bigger than it really is. In this case, we might think our sample size is ten per group, but in fact it is only one (i.e. we have no replication at all).

When thinking about replication, it is helpful to understand the concept of the experimental unit. This can be defined as the smallest piece of biological material which could, in principle, receive any of the treatments used in the study. In the example above, feeding treatments are applied to whole rats. Clearly, separate blood samples from a rat cannot receive different treatments in this study, so rat, rather than blood sample, is the experimental unit. Independent replication requires replication of experimental units. In this case, we would require blood samples taken from multiple rats fed on the two diets.

While mistakes as obvious as the one above are rare, there are more subtle ways in which non-independence can arise. Suppose we ran the experiment above with ten standard rats and ten supplemented rats. For ethical reasons, we decide to house the rats in groups of five, using four separate cages. For practical reasons, each cage is supplied with one of the diets. At first sight, this seems fine. However, since the food treatment is applied to the cage as a whole (and rats in the same cage cannot receive different food treatments), cage has replaced rat as our experimental unit. Our study initially appears to have ten independent units per treatment, but in fact it only has two. If we treated each as an independent data point in our analysis, we would be pseudoreplicating and our conclusions could well be wrong.

There are many stages where non-independence can creep into a poorly designed study. If all samples of one treatment are kept on one shelf in an incubator, with samples from another treatment on a different shelf, then shelf has inadvertently become the experimental unit. Similarly, if all immune assays from treated individuals are assayed on one 96-well ELISA plate, while controls are assayed on another, the ELISA plate has become the experimental unit. Careful thought of how units are allocated at all stages of the experiment can avoid these problems (see below).

This is not to say that there are never reasons for taking multiple measures from a single experimental unit. Such an approach can be useful in improving the precision of a measurement. For example, if our oestrogen assay is noisy, taking several measures and then combining these (for example by taking a mean) will provide a more precise estimate of the individual's oestrogen level. It is only when the individual measures are used as independent data points that the problem arises.

Take home

When reading a study ask yourself, what is the experimental unit being used in this study and has a single measure been taken from each experimental unit? If multiple measures have been taken from a single unit, have the authors explained how they have dealt with this in their analysis? If not, be very cautious, especially if the degrees of freedom in statistical analysis are more than the number of experimental units.

2.3   How are the experimental units allocated to treatments?

Let us continue with the experiment described above using 20 rats, each in their own cage. An obvious decision we must make is which rats will be allocated to which treatment group. The default procedure to use in this situation is to allocate individuals at random. Individual rats will differ in all sorts of ways, and this procedure ensures that this confounding variation is randomized across our treatment groups, minimizing the risk of systematic bias.

A frequent problem in research is the confusion of true random allocation (where, for example, rats are numbered and a random number generator is used to determine which rats go to which treatments) with haphazard allocation. Imagine your rats start in a single large cage, and you grab rats one at a time (without looking!) and allocate to a treatment. At first sight, this appears random, but it is not. Suppose rats differ in aggression level; in this case, the first rat you select is likely to be one of the more aggressive rats, and similarly for the second, whereas the less aggressive rats will tend to be selected last. If rats chosen earlier are put into one treatment and those chosen later into another, the groups will differ systematically. Even more elaborate procedures, such as alternating which treatment group a rat is put into, will not guarantee that the sample is random and will leave you open to criticism.

Random allocation is not limited to the initial set-up of the experiment. Our rat cages will be placed into a rack in the animal house. We should avoid the convenient route of putting all treatment rats in the top two rows and all control rats in the bottom two rows (or some similar allocation pattern), because treatment effects may be confounded by positional effects (e.g. perhaps the top rows are warmer), leading to inadvertent pseudoreplication. Instead, cages should be randomly allocated to positions in the rack to avoid problems.

Sometimes, a researcher may, for good reason, forego complete random allocation in favour of some other strategy. In a study involving both male a female mice, completely random allocation may lead to many more females in one treatment group than another. In this case, a researcher might decide that that they will allocate half of the males and half of the females to each treatment. This procedure is called stratifying (in this case by gender), and it leads to what statisticians refer to as a randomized block experiment. However, even in this case, which males and females are allocated to each treatment should be random.

Similarly, if an experiment needs to be split between two incubators (or divided in some other way), the researcher may decide to stratify by incubator, ensuring that treatment groups are split evenly between incubators (but randomly allocating within treatment groups). However, in situations where complete randomization is not used, it is the duty of the researcher both to explain and to justify this decision.

Take home

Haphazard is not random. Be cautious of studies which do not explain explicitly how units were allocated to treatment groups and the experimental apparatus, and do not justify situations where randomization was not used.

2.4   Are controls present and appropriate?

Experimental controls provide a baseline with which to compare our results. A trial that shows that sufferers from colds who take a particular homeopathic remedy feel better the following day tells us very little. Colds generally get better anyway, and we have no idea whether the health of these individuals would have improved in the morning without treatment.

The best controls should be identical to the treatment in every way, except for the specific treatment of interest. They should be carried out at the same time, in the same place, under the same conditions, assayed in the same machine, etc. Sometimes identifying the correct control requires careful thought. If the drug we wish to test is administered by injection, dissolved in saline, then simply having a second group of individuals who do not receive treatment is not providing the appropriate control. Our treatment and control groups differ systematically in ways other than the drug (the most obvious being that one group are injected while the other is not). In this case, the appropriate control would be to inject individuals with saline. In human studies, such sham procedures are also essential for avoiding placebo effects.

Sometimes, a control is not necessary or even ethical. In a trial comparing efficacy of a new drug and an established treatment for a serious disease, it would usually be unethical to have untreated individuals as a control group. It is also unnecessary if our question is about their relative effectiveness, rather than the effectiveness of the treatments *per se* (and we are happy to assume that this has already been well demonstrated for the established treatment).

The kinds of control described above are more formally called negative controls. Some of studies also require positive controls. These are samples which are used to validate the experimental procedures. For example, suppose you test whether treating cell cultures with ultraviolet light leads to expression of a particular gene which is not expressed in the negative controls. If you find no expression in either group, this may show that the gene expression is not affected. However, an alternative possibility is that the expression 'assay' is not working, so you cannot see the difference in gene expression. The inclusion of a positive control (e.g. a cell line that always expresses the gene product) would allow the researcher to exclude the possibility that the assay procedure is not working.

In assaying experiments, special care must also be taken to control for observer bias. This is where knowledge of the treatment group being assayed, coupled with an expectation about the outcome of the experiment, unconsciously biases the measurements being made. The solution to this is simple: whenever possible, researchers should assay experiments blind (i.e. without knowledge of which treatment they are assaying).

Take home

When evaluating someone's work, ask whether controls are in place. If not, there may be serious limits to the inference that can be drawn. If control groups are present, ask whether they control appropriately for all aspects of the treatment – and, if not, what alternative interpretations might be possible. Finally, have blind procedures been used to avoid observer bias?

2.5   How appropriate are manipulations and measures as tests of the hypothesis?

Does early exposure to cigarette smoke cause increased asthma in children? Testing this directly would require manipulation of the putative causal factor (exposure of children to cigarette smoke) and measurement of the response of the factor of interest (asthma in children). Such a study is obviously unethical. Another possibility would be to carry out the same experiment using an appropriate laboratory model (a rodent, perhaps). A third possibility would be to expose tissue cultures to chemicals present in tobacco smoke and measure the expression of genes linked to asthma development.

The two latter experiments differ from the first in that the hypothesis is tested indirectly; we manipulate factors which we assume are suitable surrogates for the putative causal factor (exposure of rodents or cells to cigarette smoke) and we measure the response of other surrogate measures (rodent asthma or gene expression linked to asthma). Indirect experiments can provide important tools for addressing questions which cannot be tackled directly for practical or ethical reasons. However, relating their results to the actual hypothesis of interest requires making assumptions which may or may not be true. Ultimately, how useful the conclusions are depends on our confidence that the measures taken are really suitable surrogates for the things we actually want to know about.

Thus, care needs to be taken in interpreting studies which use surrogate measures to test hypotheses, and it is critically important to keep a clear distinction between what was actually manipulated and measured (i.e. what the experiment actually showed) and what we are really interested in. All too often, this distinction is lost after the methods are described, and results are presented and discussed as if a direct experiment had been carried out.

Take home

When evaluating someone's work, make sure you are clear about what is being hypothesized to have an effect on what, then ask whether these factors have been manipulated and measured directly, or whether surrogate measures have been used. If surrogate measures are used, does the author justify the choice of surrogate and clearly distinguish between their results and what they would like to infer from them?

2.6   Final words

There is no perfect experiment, but the better the experimental design, the stronger the inference that can be drawn about the hypothesis being addressed. Any real study will sit somewhere on a continuum from 'extremely poor' to 'extremely good', and the value of understanding design is to be able to place an experiment in its position on this continuum. It is only by understanding the quality (and limits) of the study that we can fully evaluate its results.

Further reading

For more details on all of the above, try Ruxton, G.D. & Colegrave, N. (2011). *Experimental Design for the Life Sciences*, 3rd edition. OUP.

For a more advanced treatment of the topics, including the link between statistics and design, try Clewer, A.G. & Scarisbrick, D. (2001). *Practical Statistics and Experimental Design for Plant and Crop Science*. Wiley and Sons Ltd.

3
Statistics: a Journey that Needs a Guide

Gordon Drummond

Anaesthesia and Pain Medicine, Royal Infirmary, Edinburgh, UK

Some experiments give results that are self-evident and may not need statistical analysis. However, all results that are random samples will contain at least some random variation. To judge whether random variation could be source of any observed differences in the results of our experiments, statistical analysis has to be used. Competent help with statistics is often inaccessible to researchers and authors, and the alternative sources of information on offer may deceive and delude, like a Will o' the Wisp. A qualified statistician is worth his or her weight in gold.

Basic books are frequently inappropriate and concentrate on 'classical' methods that are unsuitable, and software varies in the guidance it gives and often doesn't warn if you get off the track. Research workers continue to 'do as we have always done' which can often be wrong. All the surveys of scientific papers that have been done (and there have been many) find extensive serious statistical shortcomings (Curran-Everett & Benos, 2009). Thus, you will frequently find papers that are statistically inept, even wrong. It is clear that the inferences that authors make on the basis of flawed or mistaken statistical approaches should not carry the weight intended and the reader would be justified in being increasingly sceptical about the strength of the evidence. This primer does not pretend to give you answers; instead it suggests questions that can help you address the question, *'What should I look for when considering the statistical competence of a paper?'*

3.1 Is the design suitable?

Has a specific, answerable, question been asked? Does the experiment provide data that can answer it? A basic rule of science is that for every postulated 'fact', it should

Essential Guide to Reading Biomedical Papers: Recognising and Interpreting Best Practice, First Edition.
Edited by Phil Langton.
© 2013 by John Wiley & Sons, Ltd. Published 2013 by John Wiley & Sons, Ltd.

be possible to devise an experiment to prove it wrong. Does the paper answer the question that has actually been asked, rather than answering another question instead? Many papers start with a very woolly question: 'mechanism' is a common [1]weasel word. Is 'mechanism' the step from gene to product, or the entire control process? Do the abstract, introduction, and conclusion match up? In essence, does the study do what is 'says on the tin'?

3.2   How are the data presented: is their nature evident?

Even before data are tested, they should be inspected. Often you cannot tell what you are being shown. If the number of samples is small, there is no excuse not to show each piece of data in a dot plot. Hiding the pattern of the data inside an innocent-looking bar chart may flatter to deceive; are the authors trying to hide something ugly in their data (see Figure 3.1)?

By way of example, in Figure 3.1 below, the same data are presented as columns (dynamite plunger) and in a dot plot. In the dot plot, we can see all the values and how they are distributed. The right hand set of values have a skewed distribution and should not be analyzed using methods that assume a normal distribution (they can be transformed, for example, into log values to allow this if necessary).

What do you think the plunger indicates? Why did the researcher choose to use it? Figure 3.2 shows a likely truth: authors want their data to look good, which is not good. Beware this author's intentions!

Most often, data should be shown as mean and standard deviation (if normally distributed) or median and quartiles (if the distribution is skewed, i.e. non-normal). Ratios and '% control' values are often skewed. If the SD is large in relation to the mean, we should suspect that the data are skewed. For example, if a mean is 3.1 and the SD is 2.3, then, if the distribution were normal, there would be some negative

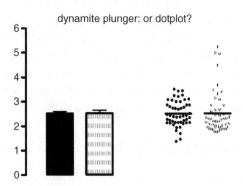

Figure 3.1 The same data represented as bars ($\pm$ SE) and as a scattergram plus the mean. Reproduced, with permission, from Show the data, don't conceal them. Drummond, G.B. And Vowler, S.L. (2011) Journal of Physiology 589.8, pp 1861–1863 © Wiley.

---

[1] 'Weasel' words: variously defined to mean things that are written or said with the intention to deceive another.

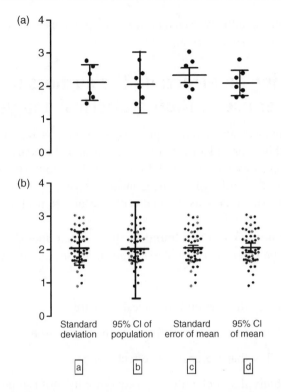

Figure 3.2   Scattergram showing different approaches to present and summarise data. Small and large samples (a) and (b) have been taken from the same population, with different ways of indicating 'scatter' (Reproduced, with permission, from Show the data, don't conceal them. Drummond, G.B. And Vowler, S.L. (2011) Journal of Physiology 589.8, pp 1861–1863 © Wiley.).

a. To describe a population's scatter, the SD (standard deviation) is a good measure. One SD each side of the mean is about 60 per cent of the values, while two SD encompasses about 90 per cent. Note that the SD is not greatly dependent on sample size.

b. In contrast, the 95 per cent confidence interval of the population tends to get bigger as the sample gets bigger, because it is increasingly likely to include the examples from either extreme of the distribution.

c. The SE (standard error) is *not* a population measure, but is a measure of the 'accuracy of estimation' of the mean. The more samples, the more precisely we can estimate the mean. It is generally *not* useful to be told the SE.

d. A preferred indication of the precision of the mean is the 95 per cent confidence interval. Most of the time, the population mean will be within this range.

values. Often, this is unlikely! Standard tests of significance are inappropriate if a population is skewed (non-normal).

3.3   Proving an effect: do the authors make proper use of their statistical analysis?

Experiments can have several consequences. They may prove something, or fail to prove something, or fail to provide a conclusion. Standard statistical logic is far from straightforward. It is based on the 'null hypothesis'. Most scientists don't understand the null hypothesis. Consider the example below, concerning a T test and the statements associated with the question (1 to 6). Try not to cheat.

You've just done a T test and found $P = 0.02$. What does this P value show?

[*The correct interpretations of, or reactions, to these statements are at the end of this primer.*]

1. The probability that the null hypothesis was true.

2. The probability that the alternative hypothesis was true.

3. The probability that the observed effect was real.

4. The probability that a claim of a positive result is a false positive claim.

5. The probability that the result can be replicated.

6. The strength of evidence in the data against the null hypothesis.

None of these results are the 'absolute truth', but the third is also a serious problem. Most studies set out to prove something, and there is a good chance that if they fail they are not submitted for publication. In itself, this is a substantial problem with the publication process.

3.3.1   The P value is an awkward concept

It means probability: 'how likely?' With the usual type of test, a positive test result is when the probability is small. This is because the possibility that the null hypothesis could be correct is *not* likely.

There are two factors involved in considering a very unlikely conclusion that 'the null hypothesis is correct' - meaning that we have 'made believe' that two samples we have analysed have been drawn, at random, from the same population. One is that the data are 'barn door', it is clear they are not from the same place at all: the other is that it's not barn door but the experiment has been big enough to answer the possibility very precisely.

3.3.2   Opinions differ about how we should interpret *P* values

A small value means no more than 'very, very, very unlikely' and a large value (say $P = 0.06$) means 'quite unlikely' (that the samples we have measured could have been drawn randomly from the same population). Automatically setting a line at 0.05 is only a convention. There is precious little difference between $P = 0.04$ and $P = 0.06$ – both are 'quite unlikely'.

We have to draw a line somewhere, and a long time ago the line was drawn at 0.05 by Fisher. However, he did argue that a P value should be used as part of a graded analysis and that other factors should be considered. Other statisticians, such as Neyman and Pearson, used a standard 0.05 value as a cut-off because they took a long view, reasoning that if we went on using 0.05 as a sort of 'industrial standard' for a series of tests, we would not often make the wrong decision (in the long run, if there were no difference in *any* of the tests, we would be wrong (false positive) one time out of 20). Far more important are two other questions:

- How important is it to make the correct decision?

- How big is the difference and is it important?

3.3.3   How important is it to make the correct decision?

If getting a positive answer (i.e. we think the null hypothesis is *un*likely to be true) could require us (individual or government) to spend a lot of money, then it makes sense to try to be absolutely certain of the result and so accept only a small P value. However, if big profits could be made for a small outlay, you might be much more willing to accept a less certain result or outcome – to accept a larger P value. Many research articles never think this far!

3.3.4   How big is the difference and is it important?

If an experiment takes many samples, then a very small (real) effect may become statistically evident. Is this important? If a gene is knocked out, do you only expect a tiny difference? Is a five per cent inhibition of growth by a new antibiotic important? In considering this aspect, the fact that the P value is small (smaller than 0.05) is less important than the size of the effect. Both of these considerations relate to signal and noise. A big signal should be easy to detect. On the other hand, a small signal will only show up with a precise, powerful (large) study. Large (and expensive?) studies should only be conducted if knowledge of the answer is important.

3.3.5 Do the authors confuse 'absence of evidence with evidence of absence'?

This question is at once deeply philosophical and of great practical importance. The often cited phrase '*absence of evidence is NOT evidence of absence*' is an example of an aphorism that belies a painful truth in the reporting of science that is often overlooked, sometimes deliberately. Typically, if $P = 0.06$, we may not discard the null hypothesis, even though we know that in reality the null hypothesis was not really correct. How does this happen? In very simple terms, the '[2]*signal is there*' but it is hidden by the [3]noise.

In many cases, this can be considered before experiments are performed, and such studies can make a proud claim: if an effect greater than a specified size truly exists, then the design of the study is such that it is likely to find it. In many cases, it is possible to consider the '*power*' of the study, which is to say the likelihood that the study will demonstrate a real effect of a certain size. There are some now familiar questions that associate with this overarching question:

- What size of effect is expected?

- How important is it to have a positive conclusion?

- How much variation is in the data?

Simple software can indicate the number of measurements needed (see www.surveysystem.com/sscalc.htm). It is easy to imagine situations in which it would be unethical to neglect to estimate the required sample size:

- Very costly public surveys

- Experiments that require the use of experimental animals

- Experiments that carry risks to the health of the subjects

After the experiment, pay no attention to the writer who states 'if we had made twice the number of measurements, we would have had a significant result'. This is rubbish. If you did another experiment, you would have to take your chance with random variation all over again, and the chance of getting exactly the same pattern would be very small! The important fact is that P can be a little more or a little less, and it means the same thing – it's on a scale, not a cliff.

Many papers make the following assertion: '*we found no difference, so there must be no difference*'. Most often this is a [4]*non sequitur* because the study does not have the power to find a difference. This does *not* mean there is no effect, unless the

---

[2] The effect you have reason to believe is real and which your experiment was conducted to demonstrate.
[3] The variation in the measurements.
[4] Non sequitur: Latin phrase that means "does not follow".

authors can say '*our data show that we would have been likely to find a result of such and such a size, and we didn't*'.

The error is compounded when there are two treatments, and A causes a significant difference while B does not. This takes us on to the next key question.

3.4   Is the experiment designed properly?

We are comparing the effects of A with the effects of B.

	Control	Treated	
Treatment A	*P*	*Q*	*Q* vs *P*, *P* = 0.05
Treatment B	*R*	*S*	*S* vs *R*, *P* = 0.06

It is wrong to conclude that treatment A worked better than treatment B. The proper comparison is between Q and S. In many studies, a better comparison would be between changes, e.g. all the $(Q - P)$ values and the $(S - R)$ values, that is, a paired study, assuming we could take two measures of each individual – before and after treatment. This reduces one of the major sources of variance: the variance between subjects.

3.5   How many groups are being compared?

If one is comparing more than two groups or more than two conditions, the usual test is analysis of variance (ANOVA). ANOVA tests come in two flavours:

- **One-way**: this asks, 'Is the difference *between* the groups greater than might be expected, judging from the differences seen in the values *within* the groups'.

- **Two-way**: this asks a similar question, but in this case variation is recognized to be attributable to two or more factors that distinguish the groups, as well as random (sometimes called residual) variation. For example, we could consider that both sex and strain might affect the measurement we are making in mice.

Proper reporting of ANOVA should give the degrees of freedom used in the analysis, so you can be sure how many samples were used. This is even more important when Multiple-way Analysis of Variance (MANOVA) is used. Some comparisons may be impressive but irrelevant. Proper reporting should also state the main effects and if there are any significant interactions.

Lastly and importantly, if there are differences, ANOVA does not show where these are. For this it is necessary to use 'post tests' to pick out differences, and these depend on what is being inspected:

- Only between the control and the treatments.

- Between all treatments.

There are different 'post tests' for these different comparisons.

3.6 Does the study relate to rare events?

Consider the following. You are reading a report of a new drug treatment that has been trialled. You read that there were no deaths when the drug was given to 20 healthy volunteers and that the author concludes that, 'the treatment is safe'. Suppose there was a real risk of death and that it was 5 per cent. This means that if there had been 100 subjects, there would be likely have been some deaths. With a 5 per cent chance of death and 100 subjects, as many as 16 could die. In our sample of 20, the 95 per cent confidence limits have the range from 0 to about 3. Thus, to detect rare(ish) events reliably, a large study is needed.

3.7 What *are* the data, and how are they distributed?

It can be easy to apply the wrong test. Measurements can be *categorical*. Categorical data comes in several flavours:

- **Nominal** – meaning that data belongs in one named group or another, e.g. alive or dead; red or yellow or blue.

- **Ordinal** – meaning they can be ordered or ranked, e.g. large, medium or small (it is important to recognize these sort of data, particularly if they have been given numbers, e.g. grade 1, 2, 3, which may be ordered. However, they should not generally be mathematically manipulated, such as averaged.)

On the other hand, one can have *interval data* that consist of continuous measures when there is an equal difference between successive measures, e.g. degrees of temperature. Finally, we have *ratio data*. Ratio data have a logical zero that means 'completely absent' (this would not be true of temperature in degrees Celsius, but it is true of temperature in degrees Kelvin.). Different statistical tests should be used for categorical and interval data. Additionally, the 'classic' tests, such as Student's test, should *be used only for normally distributed data*.

A great deal of laboratory data are *not* normally distributed. For example:

- ratios

- gel densities

- change from baseline.

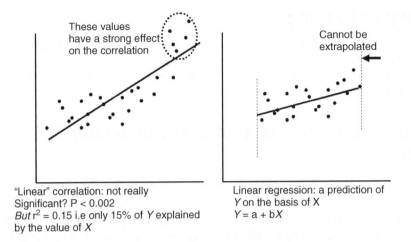

Figure 3.3   Scattergrams showing the potential limitations of linear correlation and regression. **A** The data points in the ringed area have a powerful effect on the apparent correlation but lead to a low r^2 – this data should probably *not* be regarded as suitable for linear correlation. **B** Linear regression can define the mathematical relation that explains the apparent relation between X and Y but it does not in itself show any causal relation. A regression line should not be extrapolated beyond the extent of the data.

If this does not make intuitive sense, consider 'change from baseline'. Changes from baseline may easily double or treble but are unlikely to decrease by more than 50 per cent. This means that the degree of change is not symmetrical and a normal distribution is, amongst other things, symmetrical.

Correlation is one of the most widely misused tests. Probably the most frequent of many errors is fitting a straight line to data that, when plotted, quite clearly show that the data do not conform to a straight line (most biological relationships are not linear).

Many correlations are significant, at least statistically, because of outlying values. This doesn't mean 'important'. A better measure is the r^2 value which is the proportion of the variation in one value that is attributable to the other and often expressed the '% explained'. If P is < 0.001, but r^2 is 0.2, that's NOT impressive: look very hard at the graphs and consider the outliers (see Figure 3.3).

Linear regression is *not* the same as correlation. Regression is a way of predicting one value from another. Typically, one value is considered fixed, or independent, and this determines the value of the dependent variable. In most cases, the independent variable is manipulated in the experiment.

A fundamental test to apply to the statistics you find in any paper is to ask: what number of values have been tested here, and what was the hypothesis? If you cannot tell, then you can reasonably doubt the conclusion. Beware of multiple tests done on the same sample, and never believe a subgroup analysis that has not been defined before the data were analyzed.

Further Reading

A simple, non-threatening series can be found at: http://jp.physoc.org/cgi/collection/stats_reporting

Curran-Everett, D. & Benos, D.J. (2009). Statistics, authors, and reviewers: the heart of the matter. *Advances in Physiology Education* **33**, 80.

Answers to the question posed in Section 3.3 above

You have probably guessed that the answer is none of these options!

- A *P* value is the likelihood that that you would have found these data, or data more extreme, if you had taken two random samples from a single population.

- A good expression is 'strangeness'. It would have been odd to get these data if there was no difference in the *origin* of samples (not in the samples themselves – it would be very peculiar to get two samples that were the same).

Section B
Cell and Molecular

4
Organ Bath Pharmacology

Emma Robinson
Physiology & Pharmacology, University of Bristol, UK

4.1   Basic 'how-to-do' and 'why-do' section

The term 'organ bath' is used to refer to a wide variety of apparatus that are designed to:

1. maintain a living piece of isolated tissue *in vitro*; and

2. facilitate the measurement of a response that is of interest to the pharmacologist, physiologist or biochemist.

The isolated tissue is mounted in the organ bath and is maintained under near physiological conditions. Using this approach, an isolated piece of tissue can be kept alive for many hours, allowing for detailed investigations into how substances added to the tissue influence functional responses (see also Primer 5).

Originally, organ baths were developed to study classical excitable tissue such as axons and muscles, but they have been adapted to study epithelia and all manner of intercellular communication. The most common use for the organ bath is to study the actions of drugs *in vitro* to characterize their properties as [1]agonists or [2]antagonists. Although organ bath approaches can be applied to a variety of mammalian and non-mammalian tissues and organs, the most widely used approach utilizes muscle contraction as the functional measure. Studies investigating neuronal function can measure the consequence of neuronal input and transmitter release through muscle contraction or by directly recording electrical activity.

---

[1] An agonist is any substance that binds to a receptor and triggers a response by that cell.
[2] An antagonist prevents the binding of an agonist at a receptor molecule, inhibiting the agonist-induced response.

Essential Guide to Reading Biomedical Papers: Recognising and Interpreting Best Practice, First Edition. Edited by Phil Langton.
© 2013 by John Wiley & Sons, Ltd. Published 2013 by John Wiley & Sons, Ltd.

It is now typical for organ bath experiments to combine several techniques in order to make more powerful observations. Examples include contractile force with intracellular membrane potential (Primer 9) or membrane current (Primer 10), or contractile force with intracellular ion activity (Primer 17), etc.

The response of interest is typically recorded using a 'transducer' that converts the biological signal (force, transmitter release or intracellular calcium release) into a electrical signal (voltage, resistance or capacitance) that can be digitized and recorded by a computer. The experimental design (see Primer 2) will generally seek to establish a correlation between the magnitude of a response and changes in intracellular signalling processes that are believed to underlie the response. Depending on the tissue type selected, different physiological and pharmacological targets can be investigated (see Table 4.1). These may be specific receptors, which are expressed on the muscle cells and intracellular targets, or receptors associated with neuronal innervations to the tissue, or responses to environmental factors such as stretch or pH or partial pressures of respiratory gases such as oxygen and carbon dioxide.

In pharmacological studies, a range of concentrations of the drug of interest are tested alone or in combination with another drug treatment and the data analysed to generate concentration-response curves. When planning an experiment using an organ bath, a number of key considerations must be addressed:

1. Choice of tissue, physiological salt solution and experimental equipment.

2. Drugs (reference or under investigation) and the concentration range to use.

3. Specific experimental objective, e.g. agonist potency, antagonist affinity, receptor profile of the tissue, pharmacological profile of a novel compound, bioassay. For example:

   a) An experiment to compare the agonist potency of different drugs acting at muscarinic acetylcholine receptors would investigate a range of concentration in a tissue such as the isolated ileum preparation.

   b) An experiment to investigate the function of an ion channel associated with an action potential would use a tissue such as the isolated frog sciatic nerve.

4.2   The apparatus

A cartoon showing the components of typical organ bath apparatus is shown in Figure 4.1. The tissue, in this case a section of small bowel, is normally mounted inside a small glass chamber that contains a physiological salt solution, chosen to preserve the response(s) of interest.

Other physiological variables, such as temperature, pH and oxygenation, are controlled at levels broadly appropriate for the tissue, although the partial pressure of oxygen is typically abnormally high (see later). One end of the tissue

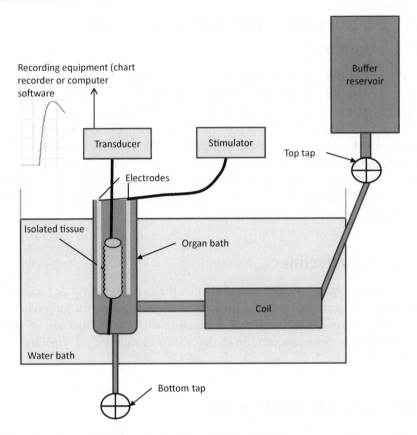

Recording equipment (chart recorder or computer software

Transducer

Stimulator

Top tap

Buffer reservoir

Electrodes

Isolated tissue

Organ bath

Water bath

Coil

Bottom tap

Figure 4.1   Diagrammatic representation of a typical organ bath designed for a single preparation. This type of equipment would commonly be found in undergraduate teaching laboratories. Courtesy of Dr. Emma Robinson.

preparation is held fixed, and the other end attached to a transducer recording either length (for a fixed load – isotonic) or force (at a fixed length – isometric). Contraction of the preparation will result in either shortening (isotonic) or an increased force (isometric) between the fixed point and the transducer. In either case, the response is converted to a proportionate electrical signal that is recorded.

Pharmacological effects can be characterized by studying how drug treatments induce contraction. If the drug of interest mediates relaxation of the muscle tissue, the experiment will involve the addition of a drug or elevated potassium concentrations (to depolarize the preparation) to induce contraction, then investigation of the degree of relaxation induced by the test compound. Isolated tissue preparations can also be studied using electrical stimulation to trigger the release of endogenous neurotransmitters contained in the nerve endings within the tissue. Drugs that modulate neurotransmission can also be tested in this way. The guinea pig ileum and vas deferens are typically used to study the parasympathetic and sympathetic neurotransmission, respectively.

4.3 Required controls and limitations

4.3.1 Handling

The dissection and handling required to 'isolate' and mount preparations for recording is potentially damaging. Prior to starting the experiment, the viability of the preparation should be tested for using an appropriate agonist at a medium effective dose (usually the $EC_{50,}$ a concentration inducing a 50 per cent maximum response for that agonist) to ensure that reproducible responses are obtained. All drugs should be added in a standardized fashion – either according to a strict schedule of timing, or when each response has equilibrated (there is no changed in the measured variable). These measures are important and should be described in the methods.

4.3.2 Slow decline

A limitation of organ bath experiments is that the tissue is slowly degrading over time, and the experimenter should provide some justification for the period over which the preparations are studied. It is good practice to see mention of regular control responses being included in the protocols to monitor the viability of the preparation.

4.3.3 Environmental factors

Another limitation is the relatively wide range of physiological salt solutions of different composition that are used; the choice is largely arbitrary. Differences in the composition of the physiological salt solutions might appear relative modest, but concentrations of key ions can significantly influence the responses. Potassium, for example, varies between 3.5 and nearly 6 mmoles per litre – values that are at the extremes of the normal range for plasma potassium concentrations.

Most physiological salt solutions include glucose as an energy source, although very often at concentrations far above those normally recorded *in vivo*. Physiological salt solutions are typically equilibrated with oxygen (100 per cent), oxygen (95 per cent) and carbon dioxide (5 per cent) mix or normal air. The level of carbon dioxide is crucial as, in some physiological salt solutions, CO_2 and bicarbonate comprise the principle buffer system for pH. In most cases, however, the gas mix used is highly hyperoxic, and some laboratories have addressed this by advocating gas mixtures containing less than 20 per cent oxygen.

4.3.4 Choice of tissue

A selection of commonly used isolated tissues and their associated physiology and pharmacology are detailed in Table 4.1 below.

Table 4.1 Examples of tissues commonly used in the organ bath set-up and their associated pharmacology.

Tissue type	Commonly used species	Selected pharmacology
Ileum (small intestine)	Guinea pig	Parasympathetic innervations (electrically evoked neurotransmitter release inhibited by opioid receptors)
Nicotinic and muscarinic acetylcholine receptors (contract)		
Histamine receptors (contract)		
Bladder	Guinea pig, rat, mouse	Muscarinic receptors (contract)
Beta-adrenoceptors (relax)		
Aorta	Guinea pig, rat, mouse (difficult)	Muscarinic receptors (contract)
Beta-adrenoceptors (relax)		
Alpha1-adrenoceptors (contract)		
5-HT receptors (contract)		
Heart	Guinea pig, rat, mouse	Muscarinic receptors (slow force and rate)
Beta-adrenoceptors (increase force and rate)		
Uterus	Rat	Muscarinic receptors (contract)
Beta-adrenoceptors (relax)		
Histamine receptors (relax)		
Vas deferens	Rat, mouse	Sympathetic innervations (electrically evoked neurotransmitter release inhibited by alpha 2-adrenceptors)
Alpha1-adrenoceptor (contracts)		
Tracheal spiral	Guinea pig, rat (difficult)	Muscarinic acetylcholine (contracts)
Beta-adrenoceptor (relaxes)		
Gastric fundus (part of the stomach)	Rat	Muscarinic receptors (contracts)
5-HT receptors (contracts)		
Isolated nerve	Frog	Na+ and K+ channels (local anaesthetics block action potential)

There is a diverse range of tissue preparations that can be studied *in vitro* using organ bath experiments. The majority of organs in the body contain muscle cells which contract as part of their normally physiology. Specific preparations have become associated with a certain type of physiological or pharmacological investigation. Preparations are selected on the basis of whether they express the receptor or response of interest and the ease with which other techniques can be used in combination. The species used will also be an important consideration, as the degree of homology with human physiology varies between species.

There are also species-specific advantages and disadvantages associated with the same tissue type. For example, the guinea pig ileum is an excellent tissue to study the actions of drugs targeting the parasympathetic nervous system and gastro-intestinal physiology and pharmacology. In contrast, the rat ileum is difficult to use in the isolated set-up as it contracts spontaneously, interfering with drug-induced effects.

Part of the decision about which tissue to use may be whether to provide exogenous electrical stimulation. When the tissue is removed from the animal, the neuronal input from the central nervous system (CNS) and periphery is severed, but the nerve terminals remain and neurotransmitter release can be elicited by applying electrical stimulation. In the case of the parasympathetic nervous system, the ganglion is also present in some isolated tissue preparations, e.g. isolated guinea pig ileum. Another commonly used stimulated tissue preparation is the vas deferens, which provides a useful tissue for studying the sympathetic nervous system.

4.4   Experimental design

Examples of concentration response data are shown in Figure 4.2. These examples show experiments examining the contraction of tissue or relaxation of pre-con-tracted tissue. Pre-contraction can be achieved using methods such as high potassium or an agonist which induces contraction.

4.4.1   Incremental curve

This method applies each drug concentration in isolation with the drug removed using wash steps, so concentrations can be added in any order. This type of protocol is chosen for responses that are a prone to desensitization and so are not suitable for cumulative curve protocols (below).

4.4.2   Cumulative curve

This method uses sequential applications of the drug without a wash step. It can be performed over a shortened time frame but only allows for increasing applications over a pre-determined concentration range.

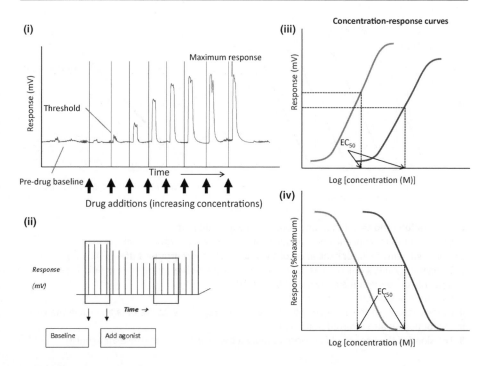

Figure 4.2 Concentration-response data illustrating agonist potency analysis in unstimulated and electrically stimulated tissue. The results shown in panels (i) and (ii) illustrate the type of output data which is obtained from an isolated tissue preparation in the stimulated and unstimulated set-up.

Panel (i) shows data for an incremental concentration response experiment recorded using a computer software program. Note how the two highest concentrations induce a similar level of response, indicating the maximum has been achieved. Increasing the concentration further is likely to desensitize the tissue.

Panel (ii) illustrates the type of data which can be recorded using the stimulated set-up. Electrical stimulation triggers the release of transmitter and contraction. For a given level of electrical stimulation, a stable response level is achieved. Drugs which inhibit the release of the transmitter can then be added and their effect observed as a reduction in the magnitude of the contraction.

Panel (iii) and (iv) show the results for full concentration response studies comparing two different agonists which induce contraction (panel iii) or relaxation (panel iv). Both types of data can be fit to a sigmoidal concentration-response curve and an EC50 value for each agonist obtained. Where more than one agonist is tested, as illustrated, their relative potency can be assessed. Drugs which are more potent sit to the left of the graph and achieve an EC50 at a lower concentration. Data may be plotted using an absolute value (panel iii) or normalised data (panel iv). Courtesy of Dr. Emma Robinson. *A full colour version of this figure appears in the colour plate section.*

4.4.3   Agonist and antagonist experiments

An agonist response curve is initially constructed using one of the methods above, then repeated in the presence of a given antagonist concentration. For incremental curves, the antagonist is added to the wash buffer so that the tissue is constantly

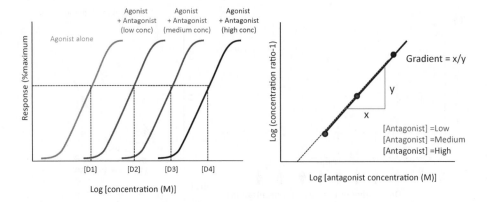

Concentration ratio calculation and constructing a Schild plot

1. Obtain values for concentrations inducing a 50% response for the agonist alone and in combination with each concentration of the antagonist concentrations ([D1], [D2] etc).
2. Using the agonist alone data as your reference, calculate the concentration ratio for each antagonist concentration e.g. D2/D1, D3/D1, D4/D1.
3. Use these data to construct a Schild plot.
4. Estimate the K_B of the antagonist by fitting a linear regression to the data and using the x-axis intercept (log KB also referred to as pA_2).
5. The slope of the line should be close to unity if the drug is a competitive antagonist.

Figure 4.3 Competitive antagonist data and Schild plot analysis. The results shown in the left hand panel illustrate concentration-response data for an agonist alone and in the presence of a competitive antagonist at increasing concentration. The rightward parallel shift in the agonist curve without a change in the maximum response indicates competitive antagonism, so therefore these data can be further analyzed using a Schild plot and an estimate of the antagonist affinity made (this is usually referred to as the K_B, with the B reflecting the fact this is an estimate of the equilibrium dissociation constant. Details of the method used to construct a Schild plot are given with an example of the linear regression analysis used to estimate the K_B given in the right hand panel. Courtesy of Dr. Emma Robinson.

exposed to the drug. For cumulative curves, the antagonist is added to the organ bath prior to the agonist. A justified amount of time must be allowed for the tissue to equilibrate with the antagonist before repeating the agonist curve. This agonist curve can be repeated for several antagonist concentrations and a Schild plot used to calculate the antagonist affinity (Figure 4.3).

Isolated tissue preparations often have significant variability between experiments, so careful experiment design is important. Where direct comparisons between different agonists or agonist responses to antagonist are made, data are ideally collected from the same tissue. It may also be necessary to construct a standard agonist curve for each experiment and normalize the data to this response.

Control experiments are required for drugs, using the vehicle in which the drug is dissolved. Both temperature and time can affect the responses obtained. For a cumulative curve, the effect of time is more significant because the tissue is not washed between drug additions and a build-up of metabolites will develop in the organ bath and affect the response. The calibration of the system should also be considered in the experiment design.

4.4.4   What data can be obtained from concentration-response curves using isolated tissue preparations?

The majority of data obtained from organ bath pharmacology are analyzed using concentration-response curves. This is a non-linear regression analysis using the Hill equation. Data are plotted using a log concentration versus response curve, and data are fitted to a sigmoidal curve. Further details relating to curve fitting can be found on the Graphpad website (www.graphpad.com). The following data are obtained from concentration-response curves:

- Maximum and minimum effective concentration.

- Potency.

- EC50 (concentration, inducing 50 per cent max response for a given agonist).

- Hill-slope.

- Antagonist affinity (Schild plot or Schild equation).

- Agonist efficacy, e.g. partial or full agonist.

4.5   Common problems and pitfalls in execution or interpretation

In general, organ bath pharmacology produces reliable and reproducible data once the optimum protocol has been developed.

4.5.1   Fitting to a model

The authors should have recorded sufficient data points to allow a reliable curve fit. If the minimum and maximum responses are not clearly defined, the curve fit and resulting EC_{50} data may be inaccurate. The Hill slope of the curve fit can be set to unity or can be a variable within the curve fit parameters. If the data lack a clear maximum or minimum response, constants can also be used, although care must be taken to ensure that this does not result in a curve fit which significantly deviates from the model. Most software used to analyze these data include statistical measures such as 'goodness of fit'.

4.5.2 Over-interpretation

Over-interpretation is a common problem. Consider this example: if the measured variable is contraction and the intervention is intended to affect changes in, say, calcium handling, there is a danger of making the assumption that observed effects are evidence of a direct association between calcium handling and force production. In such cases, a direct measure of intracellular calcium would be more appropriate, as there is good evidence that the relation between intracellular calcium and force in muscle can be significantly altered.

4.5.3 EC_{50}, IC_{50} and K_d

These can be confused. The K_d is the dissociation constant for the interaction between a receptor and a ligand. The K_d has units of concentration and can be defined as the concentration at which half the binding sites are bound by the ligand and half are free (available). The K_d can only be estimated for an antagonist using the isolated tissue set-up. The only direct method for quantifying receptor occupancy is radioligand binding studies.

IC_{50} is importantly different; it is the concentration that inhibits a response by 50 per cent. If a substance stimulates a response including agonist-induced relaxation, it will have an EC_{50} value. Why the difference between EC_{50}, IC_{50} and K_d? Consider that it may only require 10 per cent of receptors to be occupied to elicit a maximal tissue/cell response. Blocking 50 per cent of receptors may have little effect on the agonist response; indeed, one would have to apply many times the Kd concentration to observe a 50 per cent inhibition.

Concentration-response curves for agonists provide a measure of potency which is a combination of both affinity (how well the drug binds to the receptor) and efficacy (how well a drug activates the receptors and associated functional response). Antagonists have zero efficacy, while agonists have efficacy, but this can vary depending on the drug used. Some drugs are referred to as partial agonists and have lower efficacy that full agonists. For a partial agonist, the highest level of response achieved is lower than that seen for a full agonist. It is not possible to determine the affinity of an agonist from the isolated tissue preparation, but some information about efficacy can be deduced when a comparative study is carried out using several agonists acting at the same receptor.

Further reading and resources

Staff of the Department of Pharmacology, University of Edinburgh (1970). *Pharmacological Experiments on Isolated Preparations*, second edition. E & S Livingstone, Edinburgh.

Kitchen, I (1984). *Textbook of* in vitro *Practical Pharmacology*. Blackwell Scientific Publications, Oxford.

5
Small Vessel Myography

Tomoko Kamishima[1] and John M Quayle[2]

[1]Cellular and Molecular Physiology, University of Liverpool, UK
[2]Physiology & Pharmacology, University of Liverpool, UK

5.1   Basic 'how-to-do' and 'why-do' section

Appropriate contraction and relaxation of arteries plays a key role in regulating blood flow. Changes in artery diameter, and hence resistance to blood flow, is achieved by the degree of contraction of smooth muscle cells in the artery wall (but see Section 5.2 below on required controls). Early arterial contraction studies were focused on either isolated large arteries such as aorta or perfusion of whole vascular beds at constant flow or constant pressure. Though valuable, the behaviour of large vessels does not necessarily reflect that of smaller arteries, and the results from whole vascular beds may be difficult to interpret. This has led to the need for development of technologies that allow investigation of small vessel contraction.

Small arteries (diameter $<200 \, \mu m$) provide the largest resistance to blood flow where the largest drop in blood pressure occurs. Researchers may refer to such vessels as 'resistance arteries', a functional rather than an anatomical name. Resistance arteries are therefore central to understanding control of blood pressure and local blood flow and are potential targets for therapeutic agents such as anti-hypertensive and anti-anginal drugs. There are two main approaches for examining contraction of resistance arteries:

1. The isometric method measures change in *force* from dissected resistance arteries while the length (diameter) is kept constant.

2. The isobaric method measures change in *diameter* while transmural pressure across the artery wall is kept constant.

Two principal companies providing equipment for these technologies are DMT (Denmark) and Living Systems Instrumentation (USA). Historically, artery

Essential Guide to Reading Biomedical Papers: Recognising and Interpreting Best Practice, First Edition.
Edited by Phil Langton.
© 2013 by John Wiley & Sons, Ltd. Published 2013 by John Wiley & Sons, Ltd.

contraction was determined isotonically, where a change in length was measured while the load on the artery segment was kept constant (e.g. using a chymograph; see Primer 4). However, isotonic experiments are not possible for resistance arteries, due to their small size.

Broadly speaking, the isometric and isobaric methods have been applied to resistance arteries to:

- characterize the substances which modulate artery contraction, i.e. vasodilators and vasoconstrictors, and the underlying cellular mechanisms involved. Myography has been central to the study of receptor and ion channel pharmacology, as well as cell signalling. For instance, it was a key technique in discovering the role of nitric oxide as a local signalling molecule (Moncada & Higgs, 2006).

- investigate the response of arteries in different vascular beds, and often of different sizes of artery within a single vascular bed, to illuminate the underlying physiology of blood flow control. Thus, the properties of a skin artery, which acts primarily to regulate body temperature, are very different from those of a coronary artery, where blood supply must be geared to meet the varying metabolic demands of cardiac tissue.

- investigate pathological conditions in which disrupted arterial responses have been implicated. These include hypertension, atherosclerosis, diabetes, septic shock and many others.

To characterize resistance arteries isometrically, the artery of choice is carefully dissected and 'mounted' into the myograph (Figure 5.1). One of the two jaws is attached to a force transducer, while the other is connected to a displacement micromanipulator. The artery will be threaded with two wires and each secured onto one of the jaws with a pair of screws (Figure 5.1). This is challenging work, as the arteries are typically about one-fifth of a mm in diameter and the wires are almost too fine to see with the naked eye. It is very easy to damage the artery during the mounting procedure.

Once mounted, the artery has to be '*normalized*'. The aim of normalization is to improve the consistency of the contractile responses of artery sections, particularly if their diameters vary. In essence, the active contraction of a muscle (skeletal, cardiac or smooth) is dependent upon the length to which it is stretched prior to contraction. In the context of a section of artery, for example, the response to a vasoconstrictor can vary markedly, depending on how much it is stretched in the myograph. Normalization sets the initial [1]starting length to that at which the maximum active force is developed.

Interestingly, it has been shown experimentally that arteries appear to operate close to this length *in vivo* (Mulvany & Halpern, 1977). Once removed from the body and the distending pressure of the blood, it is impossible to predict the length to which it should be stretched *in vitro* in order that it develops maximal active force.

---

[1] Often called 'resting' length.

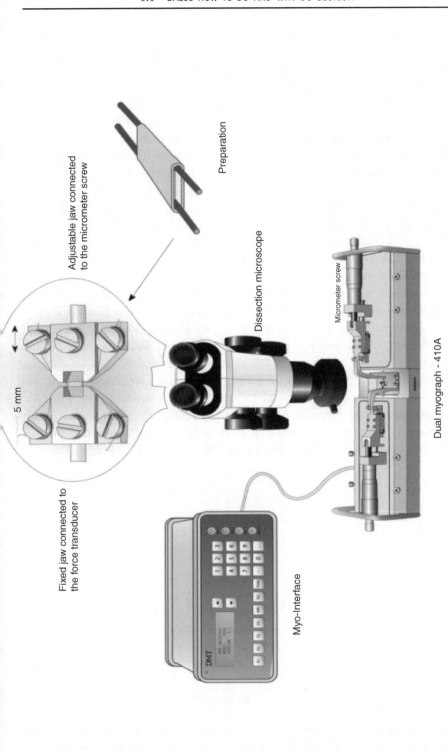

Figure 5.1 Main components of isometric wire myograph, consisting of experimental chamber, dissecting microscope and control box. Two wires are inserted to the lumen of artery ring and secured between the two jaws with a pair of screws. Reproduced with kind permission from Carsten Thorndahl © Danish Myo Technology www.dmt.dk.

Historically, a muscle preparation would be repeatedly contracted while the starting length was varied. When plotted against starting length, active force follows a bell shaped curve with the maximal level at the top of the bell. It was thus a simple task to find peak force and, from that, the optimal starting length for active force generation. This method for normalizing small arteries would take hours.

Fortunately, there is now an alternative. In 1977, Mulvany and Halpern demonstrated that the passive properties of a relaxed artery can predict the optimal starting length, and it is now common for computerized myographs to be able to perform an automated 'normalization' routine in approximately ten minutes.

In practice, a series of step-wise stretches are applied to a relaxed artery and isometric force is continuously measured (Figure 5.2). Cycles of stretch and measurement are repeated until a required reading of force is reached. It is not required that you know all the details, but it is helpful to understand the basis. You are also more likely to remember useful information accurately if you have a solid understanding of the basis.

In vivo there is a pressure difference between the inside and outside of arteries, termed the *transmural* pressure difference. If an artery's diameter is not changing, then the force that the high internal pressure generates is in balance with a tension within the wall of the artery (e.g. consider a balloon inflating). In a sense, the normalization process makes use of this relationship. Effectively, the tension at a given length is used to calculate the pressure that would generate an equal and opposite force. The target is to stretch the artery (i.e. set its length) such that the tension is equivalent to a pressure greater than 100 mmHg. The exact starting length that equates to a transmural pressure of 100 mmHg is easily determined and this length is termed 'Internal **Circumference** at **100** mmHg, or IC_{100}).

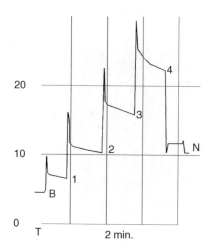

Figure 5.2 Example of normalization. The artery was stretched in a step-wise fashion until the required force reading was reached (following stretch 3), and artery diameter was then set at 90 per cent of IC_{100} (point 4) (www.dmt.dk). Reproduced with kind permission from Carsten Thorndahl © Danish Myo Technology www.dmt.dk.

Some of you may find it easier to think of the IC_{100} as the internal circumference, where the artery is fully relaxed under transmural pressure of 100 mmHg. Mulvany & Halpern (1977) found that the optimal starting length was 90 per cent of the IC_{100} value (Figure 5.2, point 4). Because all vessels are set to this standard (i.e. normalized) starting length, it allows responses of different arteries to be compared easily.

The essential property of a force transducer is that it detects force and transduces (i.e. changes) it into something else. This can be resistance, capacitance or voltage, but more often voltage. Importantly, there should be a linear relation between force and voltage, so that a doubling of force results in a doubling of voltage. Typically, this voltage signal will be converted to a digital signal (i.e. digitized) and recorded on a computer.

5.1.1   Pressure myograph (sometimes also called isobaric or flow myograph)

An important physiological role for resistance arteries in some vascular beds appears to be to render local blood flow constant over a wide range of arterial pressure. For example, cerebral blood flow has been shown to be remain remarkably steady, regardless of changes in blood pressure (e.g. during a change in posture). In such vascular beds, blood pressure itself appears to an intrinsic regulator of artery diameter; arterial contraction is, in effect, pressure-sensitive. The isobaric or 'pressure' myograph was developed in order to study the physiology of pressure-induced contraction of small arteries, a response that has become known as the *myogenic response* (Osol & Halpern, 1985).

To enable isobaric measurements, the artery segment has to be dissected cleanly from surrounding fat and connective tissue and then cannulated on a small glass pipette at one end (Osol & Halpern, 1985). It is usually secured using a single strand teased out from a surgical silk suture. The blood inside the artery is then gently flushed out of the lumen with a physiological saline solution down the cannula and the other end of the artery is tied off with suture or attached to a second cannula (Figure 5.3). The transmural pressure can then be regulated using one of two methods:

1. **Pressure head:** This requires that reservoirs of physiological saline solution are moved up and down as required to alter the vertical height of the column of saline and, in this way, set the hydrostatic pressure at the level of the artery in the myograph. A pressure transducer at the level of the artery is typically used to provide a measure of the pressure.

2. **Servo-controlled pump:** This relies on a small pump to develop the required pressure. In this case, a pressure transducer is used not only to provide a measure of pressure for the experimental record, but also to provide the input value to a control circuit that the experimenter uses to set a 'command' pressure. The circuitry uses the principle of negative feedback to reduce the

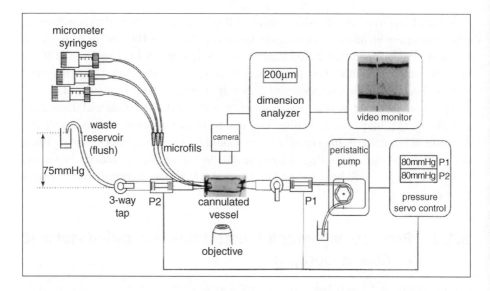

Figure 5.3 Pressurized resistance artery myograph. The vessel is cannulated at either end on glass cannulae. The artery is imaged on a video monitor and diameter is assessed by contrast detection of the arterial wall in the dimension analyzer. Transmural pressure is controlled by a servo pump connected to one of the cannulae. In the configuration shown, substances can be delivered to the lumen of the vessel via the cannulae, and to the outside of the vessel by perfusing the organ bath. This allows the effects of the application of substances separately to endothelial and smooth muscle cells to be distinguished. Figure 1 in Charybdotoxin and apamin block EDHF in rat mesenteric artery if selectively applied to the endothelium. Joanne M. Doughty, Frances Plane, and Philip D. Langton. (1999). American Journal of Physiology - Heart vol. 276 no. 3 H1107-H1112.

difference between the input and command values. In this way, the [2]pump will move saline into or out of the system until the pressure measured by the pressure transducer is equal to the command pressure.

For successful pressurization, the artery segment must be free of all holes and branches. The basic system for detecting artery diameter changes is a video system equipped with edge detection device (Figure 5.3). The magnified image of the artery is used to monitor diameter and the image contrast is adjusted to allow the detection of the artery walls. To allow the detection of artery walls, the artery segment needs to be cleanly dissected. This is technically challenging.

If the setting up of the experiment is successful, the pressurized artery should display a characteristic 'myogenic' response (Figure 5.4, McCarron *et al.,*1997).

Over the physiological range of transmural pressures, the diameter of a myogenic artery shows a characteristic 'negative slope', whereby increased pressure causes a progressive decrease in artery diameter (Figure 5.5, Osol & Halpern, 1985). The myogenic response requires extracellular calcium.

Arguably, isobaric measurement is a more physiological assessment of artery behaviour than isometric measurement, as it offers a closer approximation to the *in*

[2] Often called a 'servo-controlled' pump.

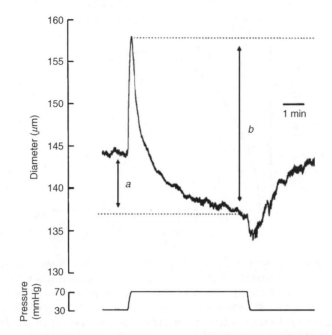

Figure 5.4 Myogenic response of a resistance artery. When pressure was increased from 30 to 70 mmHg (bottom trace), a transient increase in artery diameter occurred (top trace), reflecting the passive mechanical response to increased pressure. This was followed by a sustained diameter decrease (top trace), developing over several minutes, reflecting active smooth muscle contraction. Upon reduction of pressure back to 30 mmHg, the small transient decrease in artery diameter was followed by sustained diameter increase. Figure 1 in J G McCarron, C A Crichton, P D Langton, A MacKenzie and G L Smith (1997). Myogenic contraction by modulation of voltage-dependent calcium currents in isolated rat cerebral arteries. The Journal of Physiology, 498, 371–379. © John Wiley & Sons Ltd.

vivo environment. In addition, if the artery is cannulated at both ends, the pressurized myograph configuration provides an interesting opportunity to investigate the effects of luminal flow. Various vasodilating mechanisms are postulated to work through sheer stress (i.e. flow) acting on endothelial cells, and this type of investigation is more difficult to perform with other types of experimental equipment.

5.2   Required controls

Although smooth muscle cells are the ultimate provider of artery contraction, endothelial cells are important in regulating the contractile state of the underlying smooth muscle cells. As endothelial cells line the luminal face of all arteries, they can be easily and accidentally damaged while inserting wires – although, if you are lucky enough to be the first to make a simple observation like this, and bright enough to recognize the significance of your observation (Furchgotte & Zawadzki, 1980), then a Nobel Prize may await you (see www.nobelprize.org/nobel_prizes/medicine/laureates/1998/)!

The standard technique to assess the integrity of endothelial cells is to examine acetylcholine-induced relaxation of the pre-constricted artery. Acetylcholine is

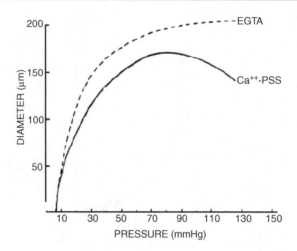

Figure 5.5 Pressure-diameter relationship of a resistance artery, determined isobarically. Note the diameter of the artery decreases at higher transmural pressure, exhibiting negative slope. When extracellular calcium is removed by chelation with EGTA, the diameter increases passively with applied pressure. Reproduced, with permission, from Figure 7 in Osol, G. and Halpern, W. (1985). Myogenic properties of cerebral vessels from normotensive and hypertensive rats. Am J. Physiol. 249 (Heart Circ. Physiol. 18): H914–H921. © American Physiological Society.

known to cause endothelium-dependent relaxation of many – but not all – arteries by several mechanisms, including release of the vasodilator nitric oxide (Angus & Wright, 2000). In some experiments, it is required to remove endothelial cells altogether to test for the involvement of the endothelium in a response. This is often achieved by mechanical disruption of endothelial cells. Commonly, a human hair is inserted into the artery lumen and, holding it between two sets of forceps, the researcher can rub the inside of the artery forwards and backwards. This procedure may damage the artery itself, and therefore it is essential to check the magnitude of artery contraction and lack of acetylcholine response afterwards.

Studies should acknowledge that some responses exhibit desensitization that will cause repeated stimuli to result in progressively smaller responses. Steps both to minimize and to monitor this, and to control for the effects on the data, should be evident in the controls.

5.3 Common problems and errors in the literature

Although the isometric and isobaric myographs are well established experimental techniques, there are some problems that can be identified in the literature. Things to watch out for include:

1. Isolated vessels are useful models, but there are significant differences between the environment experienced by an artery *in vivo* and *in vitro*. The extent to which the sympathetic nerve terminals remain active *in vitro*

is currently unknown, although it is known that it is possible to cause contraction by electrically stimulation of sympathetic nerves *in vitro*. The arteries are no longer blood-filled and the physical stresses are different. Isometric recordings tend not to exhibit resting force when adjacent artery sections mounted for isobaric recordings do exhibit resting tone.

2. There is evidence that the behaviour of vascular preparations can alter over the course of an experimental day. This has been shown to be in part the result of an inflammatory response that is initiated by exposure to non-sterile equipment, and which depressed the response to many vasoconstrictors. Very, very few laboratories conduct experiments on freshly isolated vascular tissue under sterile conditions. A good paper will consider the possibility that their experiments are influenced by an inflammatory response, or they will use pharmacological tools to depress the inflammatory response.

3. The physiological salt solutions that are used in laboratories that study acutely isolated blood vessels vary quite widely. Levels of glucose and potassium vary markedly, and both have been shown to influence the behaviour of some arterial preparations. Most laboratories also oxygenate their solutions with 95 per cent oxygen and 5 per cent carbon dioxide. The resulting partial pressure of oxygen is very much higher than that experienced *in vivo* and it is not clear that this is without effect. Increasing numbers of laboratories are starting to use a gas mixture that is more physiological.

4. There are differences between the physiology and pharmacology of different arterial beds in the same species (coronary verses mesenteric verses cerebral, etc.) and between the same beds from different species (e.g. cerebral arteries of rat and rabbit). Studies will often cite evidence from studies performed in different species or a different bed. The weight of this evidence should be considered carefully.

5. Particularly with respect to experiments using the isometric recording of force, it will be problematical to distinguish clearly where drugs are acting, because drugs that are added to the myograph chamber will have ready access to both the serosal (outer) and intimal (inner) surfaces of the blood vessel. This means that authors will need to be conservative (cautious) in their interpretation of the data collected.

6. Dose-response curves to vasoconstrictors and vasodilators are often cumulative. However, the response of the artery can also fade (or potentially desensitize) over time. Are appropriate time controls shown? In other words, does the response change over time at the same concentration of vasodilator/vasoconstrictor? This can be a particular problem when measuring relaxation, where arteries have to be pre-constricted with vasoconstricting agents. Unless the relaxing agents are fast acting (and many are slow acting),

then gradual and spontaneously occurring 'fading' of the force may prevent an accurate detection of true relaxation. This will be particularly problematical if a cumulative dose-response of relaxing agent is required.

7. In the isobaric method, once the artery has been cannulated and pressurized to a physiological pressure, intrinsic myogenic tone usually develops over a period of about an hour. Almost all resistance arteries should develop some level of myogenic tone. However, in the literature you may see that the authors have added another vasoconstrictor, such as an α-adrenergic agent, to induce contraction. These arteries may not have developed myogenic tone and may not be fully functional.

5.4 Complimentary and/or adjunct techniques

These are dealt with in the following primers of this book:

- Primer 9: Intracellular micropipette recording

- Primer 10: Single electrode voltage-clamp

- Primer 18: Fluorescent measurement of ion activity in cells.

Autonomic nerves in the vessel wall can also be stimulated to trigger neuro-transmitter release using field electrodes (Primer 28).

Further reading and resources

Moncada, S. & Higgs, E.A. (2006). The discovery of nitric oxide and its role in vascular biology. *British Journal of Pharmacology* **147**, S193–S201.

Mulvany, M.J. & Halpern, W. (1977). Contractile properties of small arterial resistance vessels in spontaneously hypertensive and normotensive rats. *Circulation Research* **41**, 19–26.

Osol, G. & Halpern, W. (1985). Myogenic properties of cerebral blood vessels from normotensive and hypertensive rats. *American Journal of Physiology* **249**, H914–H921.

Doughty, J.M., Plane, F. & Langton, P.D. (1999). Charybdotoxin and apamin block EDHF in rat mesenteric artery if selectively applied to the endothelium. *American Journal of Physiology* **276**, H1107–H1112.

McCarron, J.G., Crichton, C.A., Langton, P.D., MacKenzie, A. & Smith, G.L. (1997). Myogenic contraction by modulation of voltage-dependent calcium currents in isolated rat cerebral arteries. *Journal of Physiology* **498**, 371–379.

Furchgotte, R.F. & Zawadzki, J.V. (1980). The obligatory role of endothelial cells in the relaxation of arterial smooth muscle by acetylcholine. *Nature* **288**, 373–376.

Angus, J.A. & Wright, C.E. (2000). Techniques to study the pharmacodynamics of isolated large and small blood vessels. *Journal of Pharmacological and Toxicological Methods* **44**, 395–407.

6
Mammalian Cell Cultures: The Example of Airway Epithelial Cell Cultures for Cystic Fibrosis Research

Scott H Randell

Cell and Molecular Physiology, University of North Carolina at Chapel Hill, USA

6.1 Basic 'how-to-do' and 'why-do' section

6.1.1 What are mammalian cell cultures used for?

Since the landmark studies of George Gey that established the HeLa cell line, instrumental for poliovirus propagation and vaccine development (Scherer *et al.*, 1953), mammalian cell culture work has been a mainstay of biomedical research. *In vitro* cell culture experiments critically support many studies, including understanding the effects of specific manipulations (e.g. genetic or chemical) on cell growth, proliferation, morphology, gene expression, function, etc. There are literally thousands of publications reporting mammalian cell culture experiments every year.

High throughput screening commonly used in pharmaceutical development is typically based on convenient 96- or 384-well format cultures compatible with robotic handling, automated end point assays and other advances to enhance relevance (Mayr & Bojanic, 2009). Despite the utility and importance of mammalian cell cultures, a critical point is whether they are representative of key cell types in their native *in vivo* environment and if they are predictive of responses in the intact human or animal. In this primer, we will explore the use of cell line models and primary human airway epithelial cell cultures towards

Essential Guide to Reading Biomedical Papers: Recognising and Interpreting Best Practice, First Edition.
Edited by Phil Langton.
© 2013 by John Wiley & Sons, Ltd. Published 2013 by John Wiley & Sons, Ltd.

understanding the pathogenesis of cystic fibrosis (CF) lung disease and for the development of novel therapies.

Briefly, CF is the most common potentially fatal, life limiting, monogenic recessive genetic disease of Caucasians. Two unaffected carriers ($\approx$1:30) have a 25 per cent chance of an affected offspring ($\approx$1:3000 live births). Disease manifests in tubular organs lined by epithelia expressing *CFTR*, the mutated causative gene. Lack of functioning CFTR protein impairs Cl^- secretion through the apical cell membrane, which affects mucosal surface hydration. In the airways, this results in dehydrated, viscous mucus that is poorly cleared, providing a nidus for repeated infection that damages the lungs. Currently, the majority of CF individuals die from respiratory causes at a median age of $\approx$40 years. Key research directions have been recently reviewed (Ramsey *et al.*, 2012).

6.1.2 Cell systems for studying CF

Significant efforts have been directed towards developing cell models for CF research. These can be roughly divided into four categories:

The first category entails heterologous expression of mutant CFTR (e.g. cells expressing ΔF508 CFTR, the most common severe mutation, consisting of a deletion of phenylalanine at position 508) in cells that do not normally express the protein. Mutant CFTR has been stably expressed in NIH 3T3 fibroblasts (Berger *et al.*, 1991) or Fisher rat thyroid cells (Sheppard *et al.*, 1994). These cells are hardy, easy to grow and are amenable to high throughput administration of potential therapies followed by convenient isotopic (Venglarik *et al.*, 1990) or fluorescence (Galietta *et al.*, 2001) assays of CFTR functional restoration to measure treatment efficacy.

Since lung disease is a major cause of morbidity and mortality in CF, and pathology is focused in the airways, the second cell model category is cultured primary human airway epithelial (HAE) cells (nasal, tracheal and bronchial). Primary HAE cells have been cultured *in vitro* for many decades. Although these cells assume a nondescript squamoid character when grown on conventional tissue culture plastic, when they are grown on porous supports at an air-liquid interface (ALI), they polarize and differentiate to recapitulate the native pseudostratified mucociliary morphology found *in vivo* as well as key physiologic functions, including ion transport and mucociliary clearance (Fulcher *et al.*, 2005; Randell *et al.*, 2011). CFTR function can be assessed in the polarized cells using Ussing chambers, which is considered to be a critical milestone test before translation of potential CF therapies into clinical trials (Van Goor *et al.*, 2009, 2011).

Although primary HAE cells grown at an ALI are assumed to be highly relevant to *in vivo* conditions, there is an inertia barrier for procurement of suitable human lung specimens for cell harvest. Primary non-CF HAE cells are very costly when purchased commercially, and CF HAE cells are not generally commercially available. Thus, a third category of cell models consists of immortal and/or growth extended HAE cells. These can be spontaneously derived from human lung cancers, e.g. A549 cells and many others, as reviewed in Oie *et al.* (1996), or they can be created by purposeful introduction of potent viral oncogenes, e.g. Simian Virus 40 Early Region, Human Papilloma Virus E6/7 genes, plus or minus the catalytic subunit of telomerase (TERT). Several widely used cell lines of this type are available, including those created from CF primary cells (Kunzelmann *et al.*, 1993; Zeitlin *et al.*, 1991).

Cells immortalized with viral oncogenes often assume an aneuploid phenotype and are genetically unstable, with frequent chromosome re-arrangement events. There are examples of CF cell lines of this type that have been genetically 'corrected' by addition of wild-type CFTR with a gene therapy-type vector (Flotte *et al.*, 1992). A minority of the immortal cell types are capable of polarizing and creating ALI cultures (Zabner *et al.*, 2003), and this may be passage number dependent.

A variation on the genetic immortalization/transformation technique is the introduction of TERT, along with other genes that overcome specific senescence mechanisms (Ramirez *et al.*, 2004). Cell lines made in this manner are more likely to remain diploid and are capable of polarization and differentiation, but they grow more slowly and for fewer passages than cells transformed with potent viral oncogenes (Fulcher *et al.*, 2009).

The fourth category is the creation of cells exhibiting respiratory tract epithelial differentiation from embryonic stem (ES) or induced pluripotent (iPS) cells, which is a new and rapidly expanding field (Kadzik & Morrisey, 2012). At this time, methods have been established to convert mouse and human ES or iPS cells, including iPS cells from a CF patient, into definitive endoderm and then into progenitors capable of airway and distal lung epithelial differentiation (Longmire *et al.*, 2012; Mou *et al.*, 2012). However, more work is needed to generate uniform, functional populations capable of traditional studies, such as Ussing chambers.

6.1.3   Types of studies and end points commonly assessed

A key research goal is to find chemical substances that can improve CFTR activity in cells containing mutant versions of the gene/protein – so-called '*CFTR modifiers*'. There are different classes of CFTR mutations that accordingly require different modifiers. CF cells harbouring a mutation resulting in a premature stop codon (e.g. G542X) would theoretically benefit from treatments promoting 'read through' of the stop. It was originally noted that aminoglycoside antibiotics can serve this function, and more recently Phase 2 clinical trials report a small molecule (Ataluren, PTC Pharmaceuticals) as potentially helpful (Wilschanski *et al.*, 2011).

The most common CF mutation, ΔF508, results in a misfolded protein that is recognized as abnormal and is targeted for proteasomal degradation rather than insertion in the cell apical plasma membrane. Cells expressing ΔF508 CFTR would benefit from a 'corrector' drug that redirects the mutant protein away from the cell quality control machinery to the cell surface (Van Goor *et al.*, 2011).

Certain mutations (e.g. G551D, a missense mutation replacing the glycine at position 551 with arginine) result in a protein that is trafficked normally in the cell, but which has defective activation and/or regulation properties. Cells harbouring these mutations can benefit from 'potentiator' compounds that help activate the mutant protein (Van Goor *et al.*, 2009). Indeed, the US Food and Drug Administration recently approved a new potentiator drug (Kalydeco from Vertex Pharmaceuticals) for CF individuals with the G551D mutation.

Major research efforts are under way to discover effective modifier compounds for all classes of CFTR mutations. In general, the strategy is an initial high-throughput screen in a model cell type followed by optimization and milestone testing in primary HAE cells grown at an ALI, and then preclinical pharmacology and toxicology studies that pave the way to clinical trials.

As well as the search for CFTR modifiers, there is abundant basic research into many interrelated aspects of CFTR biology and CF lung disease pathogenesis. The approaches encompass effects of CFTR mutations on:

- CFTR folding;

- CFTR protein trafficking;

- CFTR molecular interactions;

- integrated epithelial ion transport;

- epithelial Na^+ channels, alternative Cl^- channels;

- cell proliferation, repair and apoptosis;

- mucin and mucus; mucociliary clearance;

- innate immunity; inflammation;

- interaction of the cells with viral, bacterial and fungal pathogens;

- and many others.

Some of the studies (e.g. integrated ion transport in Ussing chambers) require polarized, well-differentiated cultures on porous supports, while others can be done with cells on routine tissue culture plastic but are very difficult in the polarized cells (e.g. single channel or whole-cell patch clamp studies). The procedures and endpoints are highly dependent on the specific line of investigation. For example, severe, sustained and destructive inflammation is the hallmark of CF lung disease, and many reports compare pro-inflammatory cytokine production in CF versus non-CF HAE cells.

6.2   Required controls and common problems and pitfalls in execution or interpretation of *in vitro* experiments

In vitro mammalian cell culture experiments related to CF encompass a wide range of topics and approaches. In general, like all experiments, they require adherence to good laboratory practices and controls appropriate for the particular types of studies. Similarly, common problems and potential pitfalls likely to be encountered in mammalian cell culture experiments related to CF depend on the specific research questions, the approach, and many other potential factors. Comprehensive treatment is not possible here, but select examples are given.

6.2.1   Controls when comparing cell lines

A typical approach is to compare functional properties of CF versus non-CF cells, which can be done with primary cells from a number of different tissue donors, or with established immortal CF and non-CF cell lines. Since many immortal cell lines are genetically unstable, special care must be taken to ensure that differences observed are, indeed, a consequence of differences in CFTR expression and function rather than non-specific changes in karyotype and, consequently, other genes which have developed in the cells over time. It is reasonable to expect verification of gene and protein expression and function consistent with CF versus non-CF cell type in immortal cell line comparisons, although this is rarely performed.

Cell line instability problems can be avoided by acute CFTR gene correction in CF cells, or CFTR knockdown in non-CF cells, which can be accomplished by a variety of methods highly dependent on the cell type (immortal cell line versus primary cells) and culture conditions (plastic or ALI) being studied. For example, gene knockdown in primary HAE cells can be accomplished with electroporation of siRNA oligonucleotides (see Primer 22). In this case, good laboratory practice demands irrelevant, scrambled or mutant siRNA sequences as negative controls. Knockdown of the cognate mRNA and protein should be checked, and the gold standard for knockdown experiments is rescue by expression of a mutated, but functional, siRNA-resistant version of the knocked down protein. Similarly, experiments using viral vectors for [1]shRNA gene knockdown, or to re-express a missing protein, require appropriate empty or mutant vector controls. Added certainty can be accomplished using inducible vector systems such as 'tet-on' expression, in which cultures with or without doxycycline are compared.

Knockdown experiments studying genes essential for cell growth and differentiation typically require an inducible gene expression system to enable sufficient proliferation and/or enough time for the cells to reach the appropriate differentiation

---

[1] shRNA = small hairpin RNA.

state prior to the knockdown. Experiments employing inducible viral vectors should include cells infected with an empty vector control, or a vector expressing an unrelated gene, to assess non-specific effects of the inducing agent. Such experiments entail at least four groups, namely control and functional vectors, each plus or minus the inducing agent.

One approach for comparing function of CF versus non-CF phenotype cells is to use chemical CFTR inhibitors in non-CF cells. It is widely appreciated that chemical inhibitors potentially have non-specific effects. A vehicle control is the minimum requirement, and inactive related compounds, as well as correlative use of multiple inhibitors, may be useful to guard against non-specific inhibitor effects. Lack of an effect of the CFTR inhibitor on the parameter of interest in CFTR mutant cells lacking CFTR function is reassuring. In general, using CFTR chemical inhibitors for CF versus non-CF comparisons should be avoided.

6.2.2 Primary cell sample size and culture quality

Comparing primary cells from a representative sample of CF and non-CF individuals avoids the genomic instability and gene expression issues typically inherent in immortal cell lines, but it can introduce other problems. Primary cells from different people are likely to have multiple unique gene polymorphisms and often have different *in vitro* growth characteristics. As a result, cells from different people may be highly variable with regard to the parameter being assessed, which may change as a function of time and differentiation state, especially when studying ALI cultures with varying percentages of the characteristic cell types at different time points (Figure 6.1).

Well-designed experiments rely on *power calculations* based on the magnitude of the anticipated effect and variability in the population to determine the sample size required for there to be a specified probability (usually 0.8 or 80 per cent) that a real effect will be detected (see Primers 2 and 3). This may or may not be practical, for example when studying novel features and there is no prior basis, either in a pilot study or the literature, for power calculations. High variability and small sample size increases the chance of a type 1 error, namely finding a difference when one does not really exist ('crying wolf') or a type 2 error, when a difference exists but is not found. When studying human primary HAE cultures, it is generally necessary to examine cells from at least 4–6 donors in each group, with each measured in triplicate or greater, depending on the magnitude and consistency of the difference between CF and non-CF cultures.

Clearly, not all primary cell cultures are created equally. CF cells derive from chronically infected and inflamed lungs and there may consequently be systematic differences in cell population dynamics as a result of the *in vivo* disease process that affects *in vitro* growth capacity. It may be impossible to control for *in vivo* disease status, since no other lung disease is precisely like CF. Studies and comparisons between freshly harvested CF versus non-CF cells may thus be complicated, and an appropriate 'disease control' for CF cells (such as non-CF bronchiectasis or primary

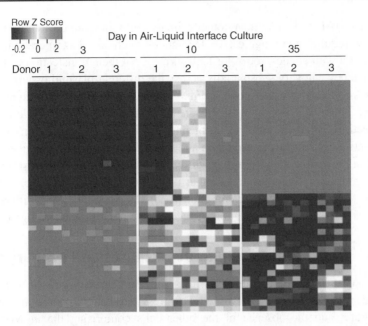

Figure 6.1 Variability in primary human bronchial epithelial cell gene expression *in vitro*. Messenger RNA was measured in quadruplicate using high throughput sequencing of ALI cultures from three different individuals at three time points as indicated. Forty of the most up- and down-regulated mRNAs are illustrated. Note that each person has a different time course and the intermediate time point is highly variable, which will introduce variability if not controlled for carefully. Also note variability in expression of down-regulated genes at the latest time point. Courtesy of Dr. Scott H. Randell. *A full colour version of this figure appears in the colour plate section*.

ciliary dyskinesia, which have a similar pattern of infection and pathology) may be difficult or impossible to obtain. Nevertheless, several studies have shown loss of the *in vivo* inflamed phenotype after more than a week of culture and with extended passage, which minimizes this concern if not studying acutely explanted cells.

6.2.3 Cell growth issues related to application of modern gene manipulation techniques in primary cells

As noted above in Section 6.2.1 ('Controls'), genetic manipulation of cultured cells with siRNA or shRNA for knockdown, or addition of a functional gene or a dominant negative counterpart, has become widely employed in cell culture experiments. Some cell lines (e.g. 293T or A549 cells) are very easily transfected with plasmids or oligonucleotides using Ca^{2+} phosphate precipitates or lipid transfection reagents. Also, in experiments employing introduction of co-expressed resistance genes and treatment of the manipulated cells with toxic selection agents to create more homogeneous populations of expressing cells, immortal cells can easily recover from the selection process.

Primary HAE cells, like many other primary cell types, are difficult to transfect, and they require more aggressive techniques such as electroporation or viral transduction with adeno-, retro-, or lenti-virus. Even then, transfection or transduction efficiency can vary substantially across the different plasmid or virus preparations, respectively. Furthermore, primary cells are mortal, with finite growth capacity. Thus, it is important that there is comparable transfection/transduction and selection efficiency across experimental groups. For example, in an experiment using lenti- or retro-virus transduction and selection for genetic manipulation, if the control and experimental viruses have different titres, selection will constrain growth of one group of cells more than another (see Primer 19). In this case, differences between the groups may be a consequence of altered growth and differentiation properties rather than manipulation of the specific gene product being studied.

6.2.4 Representativeness of cell lines and primary cells

It is important to recognize that cell lines and even primary cells *in vitro* are an imperfect representation of cells *in vivo*. The stress and artificial conditions inherent in cell culture, and the absence of the other cells comprising the *in vivo* 'niche' environment, almost undoubtedly result in different patterns of gene and protein expression, cell differentiation, metabolism, and physiologic function, etc. *in vitro* as compared to *in vivo*.

With primary HAE cells at an ALI, different choices regarding media, substratum coating and other technical factors will alter culture characteristics and performance. Specifically, regarding studies of CFTR modulator drugs, some conditions will be 'permissive', supporting relatively higher levels of CFTR gene expression and protein function in wild-type cells and 'residual' CFTR activity in CF cells (Neuberger *et al.*, 2011), while other conditions will be more 'stringent', with lower CFTR expression and activity (Randell *et al.*, 2011).

It is uncertain which of these conditions more faithfully represents *in vivo* conditions. Depending on the questions being asked, it is important to limit conclusions to those supported by the experiments. Correlative *in vivo* experiments in whole animals, examination of explanted human tissues and clinical studies in humans, when possible, are necessary to lend support to *in vivo* relevance.

Further reading and resources

Randell, S.H., Fulcher, M.L., O'Neal, W. & Olsen, J.C. (2011) 'Primary epithelial cell models for cystic fibrosis research'. *Methods In Molecular Biology* **742**, 285–310. This is the method for culturing primary HAE cells that results in 'stringent' CFTR expression conditions.

Neuberger, T., Burton, B., Clark, H. & Van Goor, F. (2011) Use of primary cultures of human bronchial epithelial cells isolated from cystic fibrosis patients for the pre-clinical testing of CFTR modulators. *Methods In Molecular Biology* **741**, 39–54. This is the method for culturing primary HAE cells that results in 'permissive' CFTR expression conditions.

Literature cited

Berger, H.A., Anderson, M.P., Gregory, R.J., Thompson, S., Howard, P.W., Maurer, R.A., Mulligan, R., Smith, A.E. & Welsh, M.J. (1991). Identification and regulation of the cystic fibrosis transmembrane conductance regulator-generated chloride channel. *Journal of Clinical Investigation* **88**, 1422–1431.

Flotte, T.R., Solow, R., Owens, R.A., Afione, S., Zeitlin, P.L. & Carter, B.J. (1992). Gene expression from adeno-associated virus vectors in airway epithelial cells. *American Journal of Respiratory Cell and Molecular Biology* **7**, (3) 349–356.

Fulcher, M.L., Gabriel, S., Burns, K.A., Yankaskas, J.R. & Randell, S.H. (2005). Well-differentiated human airway epithelial cell cultures. *Methods in Molecular Medicine* **107**, 183–206 available from: PM:15492373.

Fulcher, M.L., Gabriel, S.E., Olsen, J.C., Tatreau, J.R., Gentzsch, M., Livanos, E., Saavedra, M.T., Salmon, P. & Randell, S.H. (2009). Novel human bronchial epithelial cell lines for cystic fibrosis research. *American Journal of Physiology. Lung Cellular and Molecular Physiology* **296**, (1) L82–L91 available from: PMCID: PMC2636952.

Galietta, L.J., Haggie, P.M. & Verkman, A.S. (2001). Green fluorescent protein-based halide indicators with improved chloride and iodide affinities. *FEBS Letters* **499**, (3) 220–224 available from: PM:11423120.

Kadzik, R.S. & Morrisey, E.E. (2012). Directing lung endoderm differentiation in pluripotent stem cells. *Cell Stem Cell* **10**, (4) 355–361 available from: PM:22482501.

Kunzelmann, K., Schwiebert, E.M., Zeitlin, P.L., Kuo, W.L., Stanton, B.A. & Gruenert, D.C. (1993). An immortalized cystic fibrosis tracheal epithelial cell line homozygous for the delta F508 CFTR mutation. *American Journal of Respiratory Cell and Molecular Biology* **8**, (5) 522–529 available from: PM:7683197.

Longmire, T.A., Ikonomou, L., Hawkins, F., Christodoulou, C., Cao, Y., Jean, J.C., Kwok, L.W., Mou, H., Rajagopal, J., Shen, S.S., Dowton, A.A., Serra, M., Weiss, D.J., Green, M.D., Snoeck, H.W., Ramirez, M.I. & Kotton, D.N. (2012). Efficient derivation of purified lung and thyroid progenitors from embryonic stem cells. *Cell Stem Cell* **10**, (4) 398–411 available from: PM: 22482505.

Mayr, L.M. & Bojanic, D. (2009). Novel trends in high-throughput screening. *Current Opinion in Pharmacology* **9**, (5) 580–588 available from: PM:19775937.

Mou, H., Zhao, R., Sherwood, R., Ahfeldt, T., Lapey, A., Wain, J., Sicilian, L., Izvolsky, K., Musunuru, K., Cowan, C. & Rajagopal, J. (2012). Generation of Multipotent Lung and Airway Progenitors from Mouse ESCs and Patient-Specific Cystic Fibrosis iPSCs. *Cell Stem Cell* **10**, (4) 385–397 available from: PM: 22482504.

Neuberger, T., Burton, B., Clark, H. & Van, G.F. (2011). Use of primary cultures of human bronchial epithelial cells isolated from cystic fibrosis patients for the pre-clinical testing of CFTR modulators. *Methods in Molecular Biology* **741**, 39–54 available from: PM:21594777.

Oie, H.K., Russell, E.K., Carney, D.N. & Gazdar, A.F. (1996). Cell culture methods for the establishment of the NCI series of lung cancer cell lines. *Journal of Cellular Biochemistry* (Supplement) **24**, 24–31.

Ramirez, R.D., Sheridan, S., Girard, L., Sato, M., Kim, Y., Pollack, J., Peyton, M., Zou, Y., Kurie, J.M., DiMaio, J.M., Milchgrub, S., Smith, A.L., Souza, R.F., Gilbey, L., Zhang, X., Gandia, K., Vaughan, M.B., Wright, W.E., Gazdar, A.F., Shay, J.W. & Minna, J.D. (2004). Immortalization of human bronchial epithelial cells in the absence of viral oncoproteins. *Cancer Research* **64**, 9027–9034.

Ramsey, B.W., Banks-Schlegel, S., Accurso, F.J., Boucher, R.C., Cutting, G.R., Engelhardt, J.F., Guggino, W.B., Karp, C.L., Knowles, M.R., Kolls, J.K., LiPuma, J.J., Lynch, S., McCray, P.B., Jr., Rubenstein, R.C., Singh, P.K., Sorscher, E. & Welsh, M. (2012). Future Directions in Early Cystic Fibrosis Lung Disease Research. *American Journal of Respiratory and Critical Care Medicine* available from: PM:22312017.

Randell, S.H., Fulcher, M.L., O'Neal, W. & Olsen, J.C. (2011). Primary epithelial cell models for cystic fibrosis research. *Methods in Molecular Biology* **742**, 285–310 available from: PM:21547740.

Scherer, W.F., Syverton, J.T. & Gey, G.O. (1953). Studies on the propagation *in vitro* of poliomyelitis viruses. IV. Viral multiplication in a stable strain of human malignant epithelial cells (strain HeLa) derived from an epidermoid carcinoma of the cervix. *Journal of Experimental Medicine* **97**, (5) 695–710 available from: PM:13052828.

Sheppard, D.N., Carson, M.R., Ostedgaard, L.S., Denning, G.M. & Welsh, M.J. (1994). Expression of cystic fibrosis transmembrane conductance regulator in a model epithelium. *American Journal of Physiology* **266**, (4 Pt 1) L405–L413.

Van Goor F., Hadida, S., Grootenhuis, P.D., Burton, B., Stack, J.H., Straley, K.S., Decker, C.J., Miller, M., McCartney, J., Olson, E.R., Wine, J.J., Frizzell, R.A., Ashlock, M. & Negulescu, P.A. (2011). Correction of the F508del-CFTR protein processing defect *in vitro* by the investigational drug VX-809. *Proceedings of the National Academy of Sciences of the U.S.A.* available from: PM:21976485.

Van Goor, F., Hadida, S., Grootenhuis, P.D., Burton, B., Cao, D., Neuberger, T., Turnbull, A., Singh, A., Joubran, J., Hazlewood, A., Zhou, J., McCartney, J., Arumugam, V., Decker, C., Yang, J., Young, C., Olson, E.R., Wine, J.J., Frizzell, R.A., Ashlock, M. & Negulescu, P. (2009). Rescue of CF airway epithelial cell function *in vitro* by a CFTR potentiator, VX-770. *Proceedings of the National Academy of Sciences of the U.S.A.* **106**, (44) 18825–18830 available from: PM:19846789.

Venglarik, C.J., Bridges, R.J. & Frizzell, R.A. (1990). A simple assay for agonist-regulated Cl and K conductances in salt-secreting epithelial cells. *American Journal of Physiology* **259**, (2 Pt 1) C358–C364 available from: PM:1696431.

Wilschanski, M., Miller, L.L., Shoseyov, D., Blau, H., Rivlin, J., Aviram, M., Cohen, M., Armoni, S., Yaakov, Y., Pugatsch, T., Cohen-Cymberknoh, M., Miller, N.L., Reha, A., Northcutt, V.J., Hirawat, S., Donnelly, K., Elfring, G.L., Ajayi, T. & Kerem, E. (2011). Chronic ataluren (PTC124) treatment of nonsense mutation cystic fibrosis. *European Respiratory Journal* **38**, (1) 59–69 available from: PM:21233271.

Zabner, J., Karp, P., Seiler, M., Phillips, S.L., Mitchell, C.J., Saavedra, M., Welsh, M. & Klingelhutz, A.J. (2003). Development of cystic fibrosis and noncystic fibrosis airway cell lines. *American Journal of Physiology* **284**, L844–L854.

Zeitlin, P.L., Lu, L., Rhim, J., Cutting, G., Stetten, G., Kieffer, K.A., Craig, R. & Guggino, W.B. (1991). A cystic fibrosis bronchial epithelial cell line: Immortalization by adeno-12-SV40 infection. *American Journal of Respiratory Cell and Molecular Biology* **4**, 313–319.

7
Electron Microscopy (TEM and SEM)

Paul Verkade

Wolfson Bioimaging Facility, Physiology & Pharmacology and Biochemistry, University of Bristol, UK

7.1 Basic 'how-to-do' and 'why-do' section

7.1.1 Electron microscopy (EM)

The purpose of any microscope is to enable the observer to resolve as separate objects things that are beyond the scale of the human eye. In any microscope, the point-to-point resolution (d) is dependent on the wavelength of the illuminating source following Abbe's equation: $d = \lambda/2NA$, where NA is the Numerical Aperture, a measure of the quality of the objective – usually around 1 or slightly higher for light microscope objectives. So, in a light microscope using laser light with a wavelength (λ) of 520 nm (green) the resolution will be around $d = 200$ nm. No improvement in optical design can reduce d below this. Nowadays, however, specialized super resolution light microscopes can obtain fluorescence information below 200 nm by digital image deconvolution and reconstruction.

The electron microscope overcomes the inherent resolution limit of the light microscope by using the wavelength of electrons that are of a completely different (sub-nanometre) order. Another advantage of electron microscopy (EM) is that it will provide information on all structures present, rather than just on the location of fluorescent proteins. A disadvantage of EM, however, is that only still images can be captured, because samples have to be imaged in a vacuum. Therefore it is sometimes worthwhile to combine both LM and EM on the same sample, in one so-called Correlative Light Electron Microscopy experiment. As in the light microscope, electron microscopy can study objects inside the sample using Transmission

Essential Guide to Reading Biomedical Papers: Recognising and Interpreting Best Practice, First Edition.
Edited by Phil Langton.
© 2013 by John Wiley & Sons, Ltd. Published 2013 by John Wiley & Sons, Ltd.

Electron Microscopy (TEM) or can look at the outer surface of a sample using Scanning Electron Microscopy (SEM), analogous to a stereo light microscope.

7.1.2 Transmission electron microscopy (TEM)

The greater resolving power of electron microscopes derives from the wave properties of electrons. Unlike light waves, the wavelength of an electron varies with its speed, which in turn depends on the accelerating voltage. Standard TEMs operate in the range 10,000–120,000 V. At 60 kV, the wavelength of an electron is ≈0.05 Å (1 nm = 10 Å), but the full theoretical resolution is not realized, owing to the aberrations of electron-focusing lenses. The practical limit of resolution in the TEM is 3–5Å or 10–15Å when observing cellular structures.

Electron microscopes consist of a vacuum column through which the electrons travel. A cathode (like the metal wire in a light bulb) will be heated and provides the source of electrons. Focusing of the beam is by electromagnetic lenses. By altering the lens currents, the magnification can be varied between about 1,000–250,000 times. Electrons are brought to focus on a phosphorescent screen (which has crystals that emit visible light). Image formation depends on differential scattering of electrons within the specimen, an effect that is proportional to the sizes of atomic nuclei in the specimen (Figure 7.1).

Biological material mainly consists of atoms of low atomic number. Therefore, to increase scattering – and thus image contrast – biological material is usually treated with solutions of heavy metals during the preparation stages (Figure 7.2). These heavy metal stains specifically bind to cellular structures like membranes or DNA.

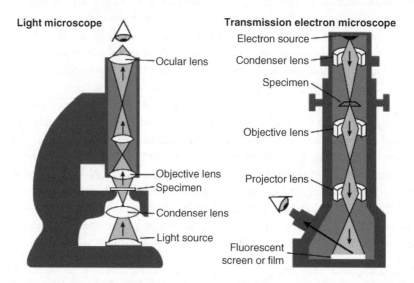

Figure 7.1 Diagrammatic representation of both the compound light (left) and Transmission Electron (right) microscopes. Blackwell Microbiology Teaching Resources (comparison of electron and light microscopes) © John Wiley & Sons Ltd.

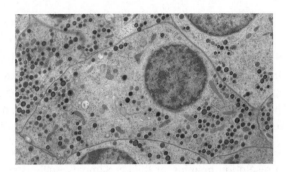

Figure 7.2   Example of a TEM image where a thin section of a piece of embedded tissue was made and imaged in the Transmission Electron Microscope. In this case, it is an insulin-producing Beta cell inside the Islet of Langerhans. It shows the nucleus as a dark grey circle and the plasma membrane is outlined by the dark lines. Inside the cytoplasm, darkly stained insulin granules can be seen. Courtesy of Dr. Paul Verkade.

The stains may, however, also limit the resolution in the EM. To record images, the phosphorescent screen is removed and a film – or more likely nowadays, a CCD – camera is brought into place.

Preparative methods for TEM

1. Most commonly, in medicine and biology, thin sections of tissue or cultured cells are cut (see Figure 7.2). The limited penetrating power of the electron beam means that ultra-thin sections (≈ 70 nm) are necessary. Preparation of these ultra-thin sections is as follows: Fixation (often by perfusion) is the first essential step, as samples will be introduced into the vacuum. This is followed by treatment with osmium tetroxide (the heavy metal contrasting agent), then by embedding in epoxy resin (a plastic), and finally sectioned on a glass or diamond knife. Sections are collected on metal mesh 'grids' and then contrasted with uranium and lead salts before they can be analyzed in the TEM.

2. To localize proteins inside cells using electron microscopy, the thin section technique can be combined with immunocytochemical staining methods, where antibodies (see Primers 12 and 13) are used to recognize specific proteins. The main requirement is for an electron-dense marker to be attached to the reaction product. Thus, gold beads are conjugated to the secondary antibody.

3. A different technique, negative staining, is suitable for examining very small particles such as protein complexes, viruses or isolated cell organelles, etc. The mesh grid is first coated with a support film (carbon or plastic). The particles of interest are attached to the support film, after which a heavy metal stain (e.g. uranyl acetate) is applied and subsequently washed off. As it dries, it leaves an electron-dense deposit over the whole surface except where the particles lie. The specimens appear bright on the screen, with their edges and surface features defined by the (very finely granular) metal salt.

4. The above techniques all use heavy metals to stain structures in the EM so we do not actually visualize the structures themselves but rather look at a 'shadow' of them. In *cryo-TEM*, we leave out the stain and image the sample directly. One requirement for this is a different means of stabilization of the sample – cryo-fixation. This can be done via plunge freezing into liquid propane. The frozen samples can then be imaged directly in an EM, where the sample is kept well frozen while imaging. This technique is able to give us the highest resolution in EM.

5. If very high resolution imaging of thicker samples, such as cells and tissue, is required, the samples will need to be cryo-fixed via high pressure freezing. The thin section technique can be also performed on frozen specimens (without fixation and embedding). Prior protection against freezing damage is usual. This method avoids denaturation of proteins, etc. It is technically very difficult to achieve.

7.1.3   Scanning electron microscopy (SEM)

Scanning electron microscopy (SEM), is very different from TEM in both construction and operation. A SEM utilizes electrons that have 'bounced off' the surface of the specimen. It is used primarily to examine the surfaces of objects, such as tissue microarchitecture or very small animals (see Figure 7.3).

The easiest specimens to examine are hard and dry; otherwise, the natural topography must be preserved, though devoid of fluid. To avoid the surface tension damage that accompanies air drying, a technique known as 'critical point drying' is used, involving the vaporization of liquid carbon dioxide under pressure. Once dried, specimens are coated with a thin layer of gold or gold-palladium and placed in the microscope.

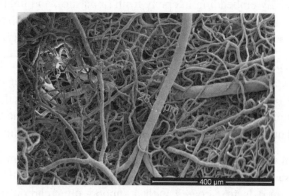

Figure 7.3 Example of a SEM image. A cast of the blood vessels in a rat brain was made using resins purchased from VasQtec, gold sputtered and imaged inside a scanning electron microscope. The image provides a 3-dimensional overview of the blood vessels (the diameter of an arteriole is shown: ≈29 microns) and shows both complexity and density of the vasculature. Courtesy of Dr. Phil Langton.

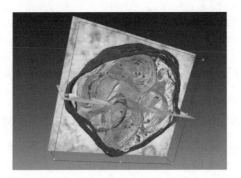

Figure 7.4 Example of a 3-D reconstruction and model of an endosome. After reconstruction of the volume, a specific slice can be analyzed. Specific features can be drawn into the reconstruction. Doing this on subsequent slices enables a 3-D model to be built. Courtesy of Dr. Paul Verkade. *A full colour version of this figure appears in the colour plate section.*

In the SEM, the electrons are accelerated as a fine beam that scans the specimen. The image is formed by back-scattered and/or secondary electrons that strike a detector near the specimen. The result is an image on the screen that corresponds to the surface features of the specimen. SEMs provide a wide range of magnification (15–150,000 ×) with a resolution of <5 nm and great depth of focus (see Figure 7.3).

7.1.4   Three-dimensional EM

While the SEM already provides some inherent three-dimensional information, there have been developments that allow for the acquisition of (better) 3-D information for both SEM and TEM samples. For the SEM, there are applications that cut off slices from the sample inside the microscope. This can be done using a diamond knife or a Focused Ion Beam but, rather than imaging the slices, the new surface is analyzed. By performing this procedure over and over again, one can zoom through this stack of images and build a 3-D model. These capabilities are absent for the TEM so, in order to obtain 3-D information, a thicker slice (300 instead of 70 nm) is made. Once in the microscope, the sample is tilted and a sequence of tilted images is acquired observing the sample from a large degree of angles, analogous to an MRI machine. The tilt series can be converted into a 3-D volume, where individual planes can be analyzed and models can be made (Figure 7.4).

7.2   Common problems or errors in literature

7.2.1   How to read EM images

A difficulty with EM is whether the images selected and published are truly representative. Because of the very high magnification, the samples chosen for study may not reflect the overall complexity of the material. Statistical methods for sampling are available, but not always practicable (for details see Mayhew & Lucocq, 2008). Also, one should realize that one or two micrographs selected for inclusion in a paper may represent a large investment of time and effort.

7.2.2   Pitfalls in execution or interpretation

Any microscopical technique based on tissue sections requires the scientist to infer three-dimensional form from two-dimensional projections. Otherwise, almost every step in electron microscopy carries the risk of various artefacts. Fortunately, these are mostly well recognized and do not mislead. Although EM observations appear to present a black and white interpretation of the data, it is important to note that (3-D) models derived from electron microscopy data are based on human interpretation of the images.

The main criticism about EM is usually that because of the fixation and processing procedures, artefacts are introduced. It is very important to note that it is because the higher resolution of the EM that these artefacts are visible. Artefacts within the LM are usually much worse, but are not recognized because they are below the resolution of the LM. There are definitely artefacts introduced by the EM procedure, and electron microscopists are generally very much aware of these artefacts; however, it takes a few years of experience to be able to recognize them and fully 'read' an EM image.

7.2.3   Before starting an EM experiment

If cells can be observed by light microscopy before and after fixation, so much the better. Any EM technique involving localization of specific proteins must have appropriate controls. Techniques that capture dynamic events should have inhibited controls.

Immuno-labelling experiments especially require a number of controls – for instance, which fixative to use, or the omission of primary antibodies (for details, see Primer 13).

7.3   Complementary and/or adjunct techniques

These are dealt with in the following primers of this book:

- Primer 8: Fluorescence microscopy

- Primer 16: GFP

- Primer 13: Immunohistochemistry

Further reading and resources (biological EM only)

Bozzola, J.J. (1999). *Electron Microscopy: Principles and Techniques for Biologists*. Jones & Bartlett.

Mayhew, T.M. & Lucocq, J.M. (2008). Developments in cell biology for quantitative immunoelectron microscopy based on thin sections: a review. *Histochemistry and Cell Biology* **130** (2): 299–313.

Maunsbach, A.B. & Afzelius, B. (1999). *Biomedical Electron Microscopy*. Academic Press. (a very advanced book about interpretation of micrographs).

8

Fluorescence Microscopy

Mark Jepson

Physiology & Pharmacology, University of Bristol, UK

8.1 Basic 'how-to-do' and 'why-do' section

8.1.1 What is fluorescence microscopy used for?

Fluorescence microscopy supports many areas of biomedical research due to the expanding use of fluorescent probes to localize proteins, DNA, lipids, ions, organelles, etc. within cells, and to identify or follow the movements of entire cells. Some fluorescent techniques reveal the static location of molecules in cells, for example using antibody labelling (immunofluorescence). Other fluorescent labels (e.g. **G**reen **F**luorescent **P**rotein (GFP) and its variants (see Primer 16) can label living cells or their components, and they are used to monitor their expression or dynamic behaviour. In most applications, it is advantageous for the fluorescent molecule (fluorophore) to be bright and stable, but the bleaching, instability, spectral switching and pH sensitivity of some fluorescent proteins can also be exploited.

Another class of fluorescent probe consist of small molecules that alter their properties (intensity or spectra) in response to changing environment or the presence of ions (Ca^{2+}, pH, membrane potential etc). Also, targeted reporters based on GFP technology have been developed to monitor conditions and enzyme activities in cells and specific cellular compartments (Primer 17).

Fluorescence microscopy is the usual end-point for experiments using any of the huge array of fluorescent probes whenever distribution between and within single cells is examined. As such, it complements other fluorescence measurement techniques, such as fluorimetry and flow cytometry, which yield information about the properties of cell populations or of single cells without reference to cell morphology or sub-cellular distribution.

Essential Guide to Reading Biomedical Papers: Recognising and Interpreting Best Practice, First Edition.
Edited by Phil Langton.
© 2013 by John Wiley & Sons, Ltd. Published 2013 by John Wiley & Sons, Ltd.

8.1.2   Introducing the range of fluorescence microscope systems

There are several types of fluorescence imaging systems, of varying capabilities and complexity. Which microscope system is used depends on the requirements of the experiment (e.g. sensitivity, resolution and speed), as well as their availability and cost. No microscope system is ideal for every application.

The simplest and cheapest fluorescence imaging systems are normal light microscopes equipped with a high-intensity light source, usually with broad-spectrum emission (e.g. Hg or Xe lamps) but, increasingly, LED and other stable, long-life, low-cost alternatives are being incorporated into microscope systems.

Light of a restricted wavelength (usually selected by a filter) is directed onto the specimen by a [1]*dichroic mirror* (beam splitter). The fluorescent light emitted from the sample at longer wavelength (and lower energy) is transmitted, by the beam splitter and a barrier (emission) filter, to a detector (either the user's eye or a camera). This type of set-up often uses a high sensitivity CCD camera to optimize signal, and it is referred to as a wide-field microscope (WFM) – referring to the fact that fluorescence is detected from a wide depth of field. WFM systems can also be expanded with filter changers, focus devices, shutters, etc. to automate imaging (e.g. for time-lapse studies).

The high sensitivity, speed and versatility of WFM systems makes them ideal for many live cell imaging studies when the best resolution is not required, when rapid imaging is required (e.g. Ca^{2+} imaging) and where photo-damage must be avoided (e.g. GFP trafficking). Images acquired at incremental levels of focus can also be enhanced by 'deconvolution' algorithms, which reassign detected light to its true point of origin by reference to its 'point-spread function' (i.e. the way a point light source is predicted to be spread by the optical properties of the microscope and specimen – see Figure 8.1). If sufficiently stable imaging conditions can be achieved, deconvolution can significantly enhance image clarity, especially improving the separation of objects at differing depths (axial resolution). However, it is best suited to imaging punctate structures in thin specimens.

Other imaging systems can provide higher axial resolution without resorting to deconvolution algorithms, although deconvolution may still enhance images from such systems. The most commonly encountered are confocal microscopes such as Confocal Laser Scanning Microscopes (CLSM) that selectively detect fluorescence from a narrow depth of focus. In CLSM, a laser beam is focused on a small point in the sample and moved through a 2-D field to build up an image point by point. The 'confocality' is generated by an aperture, placed in front of the detector, which only

---

[1] A dichroic mirror behaves in two ways, depending on the wavelength of light. For wavelengths shorter than a given value, it behaves as a mirror and reflects the light. For wavelengths longer than that value, it behaves as a pane of glass and permits transmission of the incident light. In this way, white light is 'split' with some being reflected and some passing through the mirror.

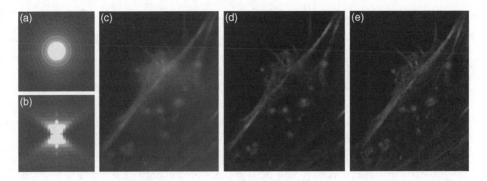

Figure 8.1 Point spread function and comparison of wide-field microscopy (WFM), WFM/deconvolution and confocal microscopy (CLSM).

Panels a and b illustrate theoretical point spread functions (PSFs), revealing how light from a point source spreads in *xy* and *xz*. Similar calculated or measured PSFs underpin all deconvolution algorithms.

Panels c to e are images of a *Salmonella* infected epithelial cell stained with TRITC-phalloidin to localize F-actin. Images were acquired from the entire cell at 100 nm focus intervals and one selected for comparison. Panel c: Raw WFM image is blurred due to out-of-focus fluorescence. Panel d: After deconvolution using Velocity™ software (and a calculated PSF), a prominent increase in clarity is observed due to removal of out-of-focus fluorescence and increased spatial resolution. Panel e: A CLSM image (acquired with Leica SP2 AOBS system) of the same cell and same plane of focus as shown in panels c and d. The CLSM image has similar level of spatial resolution to the deconvolved WFM image in panel d. Field of view approx. 20 µm × 24 µm.

Panels a and b reproduced from (http://www.bitplane.com/go/products/autoquant-x3).

Panels c–e adapted from Jepson, M.A. (2005). Confocal or widefield? A guide to selecting appropriate methods for cell imaging, in *Methods Express: Cell Imaging*, Ed D. Stephens, © Scion Press.

allows light from a single focal plane to reach the detector. 3-D images are compiled from confocal images at differing depths within the sample.

CLSM systems provide excellent resolution and facilitate the imaging of complex fluorescent distributions within thick specimens that cannot be achieved by WFM imaging, even with deconvolution, due to the excess of out-of-focus light present. However, CLSMs have comparatively low sensitivity due to a large proportion of fluorescence being 'discarded' at the aperture, and they are usually not fast, due to their point by point scanning. These limitations can be overcome to some extent by using higher sensitivity detectors and specialized fast-scanning techniques, but these improvements come at a price; thus, most laboratories are limited to more standard CLSM systems.

Alternatives to CLSM include spinning (Nipkow) disc systems, which are more sensitive and faster than standard CLSM and are well suited to live cell imaging, due to reduced photo-damage and simultaneous imaging of the entire field using a CCD camera. However, the axial resolution of spinning disc confocals is lower than CLSM, and they tend to be less flexible because they can be optimised for a relatively limited range of objective lenses.

Instead of limiting detection to one level of focus, as with CLSM and Nipkow disc systems, imaging at a single focal plane can also be achieved by limiting excitation to that plane. Multi-photon (MP) microscopy does this by exploiting the fact that the probability of more than one photon simultaneously interacting with a molecule falls off very rapidly moving away from the area of maximum photon density at the focal plane. MP excitation is possible because two photons at a particular wavelength arriving simultaneously will deliver combined energy equivalent to a single photon of roughly half their wavelength.

MP microscopes are usually equipped with lasers which deliver extremely short (femtosecond) pulses of high-intensity near infra-red (NIR) light to excite fluorophores that would normally be excited at UV or visible wavelengths. MP microscopy is used mainly to image deep within tissues, where its advantages are most pronounced due to NIR light penetrating tissue more effectively than shorter-wavelength light. On the down side, commercial MP systems are comparatively expensive and technically demanding, so are less widely available.

Since MP excitation requires high energy that can be damaging to specimens, alternative means of exciting fluorophores at single planes within 3-D specimens are being explored. One such approach which shows huge potential for imaging small organisms is Light Sheet Fluorescence Microscopy (LSFM) or Single Plane Illumination Microscopy (SPIM), in which a narrow sheet of light perpendicular to the objective lens is moved rapidly through the specimen to build up 3-D images over time. SPIM has been used to produce exceptional 3-D images of embryo development and is currently undergoing further refinement and commercialization.

Total Internal Reflection Fluorescence (TIRF) microscopy also relies on limiting the excitation of fluorophores to a shallow depth of field but, in this case, excitation is restricted to the region immediately adjacent to a glass cover slip. Light hitting the cover slip beyond the critical angle is totally reflected, but transmits some energy in a limited field (evanescent wave) adjacent to the cover slip (within 100–300 nm). This can excite fluorophore molecules that are very close to the plasma membrane of cells grown on the cover slip, or within structures attached to the cover slip in cell-free TIRF applications. TIRF microscopy can provide extremely precise localization of membrane fusion events, exocytosis, etc., but is more limited in its range of cell imaging applications than are other microscopy modes.

In recent years, several novel fluorescence imaging techniques have been developed which, in very different ways, provide 'super-resolution' capabilities, breaking the generally accepted resolution limit for light microscopy. Although these are currently far from commonplace, use of some super-resolution methods, such as Structure Illumination Microscopy (SIM), Photo-Activation Localization

Microscopy (PALM), Stimulated Emission Depletion (STED) and Stochastic Optical Reconstruction Microscopy (STORM) can be expected to expand as they are further developed and commercialized.

8.2   Required controls

8.2.1   Verifying that fluorescence signal is due to genuine labelling

Many of the controls required for fluorescence imaging studies are related to the labelling protocols rather than the use of microscopes. It is not possible to provide much detail of these here but, in the case of techniques such as immunofluorescence and FISH, the required labelling controls are described elsewhere.

8.2.2   Discrimination of multiple fluorescence signals

When two or more fluorescent labels are used together, it may be necessary to perform controls to ensure that there is no overlap in signals generated by the probes. It is also important to ensure that the fluorescence from each label is adequately discriminated by the imaging system. Most fluorophores emit (and are excited by) light over a rather broad range of wavelengths, so it is likely that if two or more fluorophores are imaged simultaneously, there will be 'bleed-through' or 'cross-talk' between the fluorophore signals that may not be immediately apparent.

To avoid this, most imaging applications will employ sequential imaging, in which each fluorophore is imaged separately with selective excitation/emission settings. This is often achieved by switching excitation filters on WFM systems or, on CLSM systems, by rapidly changing laser excitation wavelength using an AOTF (**A**cousto-**O**ptic **T**unable **F**ilter). Even when two excitation wavelengths are used sequentially, simultaneous excitation of two fluorophores by the shorter wavelength light is still common and may require stringent selection of fluorescence using appropriate emission filters.

Many papers do not provide the detailed information required to fully assess how images were acquired. In some cases, however, the possibility of bleed-through between channels can be assessed by careful examination of the images to establish whether features appearing in images of one fluorophore are absent from images of the alternative fluorophore(s). If this is the case, there must be minimal bleed-through.

8.2.3   Positive controls

As with many techniques, positive controls will verify negative microscopy results. For example, it is helpful to compare labelling of a cell type known to express a given protein to ensure that lack of labelling in a test sample is not due to failure of the labelling technique or insufficient detection sensitivity. Similarly, in dynamic studies looking for movement or change in fluorescence properties, a suitable positive control may be a treatment known to result in a detectable change.

8.3 Common problems and pitfalls in execution or interpretation

The potential pitfalls and problems encountered in fluorescence imaging studies depend on the research questions and the complexity of data interpretation. Since the range of applications of fluorescence imaging is broad, it is only possible to summarize a few issues that commonly occur.

8.3.1 Is the fluorescence signal genuine?

In studies using fluorescence imaging to demonstrate if a protein is expressed in a cell or compartment (e.g. GFP expression, immunofluorescence), proper controls should be performed to ensure the signal is not due to non-specific labelling, autofluorescence, etc. This is a vital consideration when performing the experiments, and it is surprising how often researchers mistake autofluorescence for genuine labelling. However, these basic controls are often not described in research papers, and the resulting images are very rarely shown, due to space limitations. Thus, a lack of presented negative controls should not itself raise questions about the validity of a paper.

8.3.2 Detection sensitivity

Where negative results are obtained (e.g. apparent lack of expression of a protein or gene), it is important to consider the detection sensitivity of the imaging system. Assuming that positive controls were successful, might a negative result be explained by a low level of labelling being below the detection threshold?

It is important to consider whether the sensitivity of the fluorescence imaging system used is most appropriate for the experiment. For example, CLSM is undoubtedly an excellent choice when resolution needs to be optimized, but it is not the most sensitive type of imaging system. Nevertheless, CLSM is often used when a WFM system would be a more appropriate choice and, furthermore, CLSMs are often set up to optimize resolution over sensitivity, even when greater consideration of sensitivity should be dictated by the experimental requirements.

8.3.3 Is targeting of a fluorescent probe correct?

There are many cases in the literature where the localization of a fluorescent probe has turned out to be not as specific as first thought. This can arise when a probe has a biochemically well-defined targeting domain but may have additional properties that cause it to be localized to alternative, or additional, targets within cells. It may also be the case that a probe may target a compartment but not label all of that compartment. These possibilities are always worth considering when researchers use a probe to identify a cellular compartment and compare this to the distribution of their protein of interest.

It is also possible for processing of cells to alter the distribution of fluorescent probes or protein labelling. Fixation of cells is often a source of artefacts. For example, paraformaldehyde fixation usually reveals the true distribution of proteins, but can sometimes lead to loss of antibody detection of proteins, and occasionally it has been associated with altered distribution of peptides. Similarly, solvent fixation may adversely affect localization of some proteins. Live cell imaging can often avoid these potential pitfalls by revealing protein distribution without the need for any processing. As discussed in the GFP primer (Primer 16), most GFP-labelled proteins behave similarly to the native protein. However, this is not always true, and their 'correct' localization and function should be verified wherever possible.

8.3.4 Are sufficient details provided to allow proper interpretation of imaging data?

Often, insufficient information is provided in research papers to enable full understanding of how imaging was performed. Details of optical characteristics of lenses and filters are often omitted, and this can limit data interpretation. There is also a worrying trend towards reduction in experimental detail; often, the papers do not even report the imaging system used (e.g. confocal or wide-field). While these issues may not seriously affect image interpretation for relatively simple fluorescence studies, they should be considered where there is a greater level of data interpretation. For example, the limits of resolution (determined by lens properties and other optical parameters of the microscope) and ability to differentiate fluorophores (dependent on precise excitation and emission wavelengths examined) have major implications for interpretation of apparent co-localization (see below).

A related issue is the manipulation of data. Are images truly representative, or have regions been selected and/or their intensities altered to highlight one feature while removing others? Increasingly, journals are requiring authors to assert, or even prove, that inappropriate image manipulation has not been applied and, in some cases, to provide access to raw data image data and a full explanation of how they are processed and quantified.

8.3.5 Interpreting apparent co-localization of fluorophores

Analysis of the co-localization of fluorophores requires their signals to be discriminated effectively, and for any displacement between signals arising from chromatic aberration or filter alignment to be eliminated or corrected for. In live cell applications, it is also important to image fluorescent probes simultaneously, or to consider possible temporal displacement of signals arising from sequential image acquisition.

Where two or more fluorescent labels appear to be partially co-localized, this should be adequately quantified (there are a number of algorithms for such analysis) and, whatever the result, this should not be over-interpreted as evidence that, for example, two proteins interact within the cell. Co-localization of fluorescent molecules at the light microscopical level is only evidence that the two are within the resolution level of the imaging system, so it is crucial to know (and report) sufficient details to assess the resolution of presented images.

If the highest specification optical elements are used, the lateral resolution limit of conventional light microscopes is around 150–200 nm and the axial resolution is, at best, about 300–400 nm (if CLSM or MP are used) – distances which are larger than many organelles. The resolution is inferior if low magnification lenses are used. The size of pixels in acquired images is also a factor to be considered, since images often have a pixel density that is too low to allow optimal lateral resolution to be achieved – for example, because the camera used has larger pixels or the area scanned in CLSM is too large to achieve this.

8.3.6   Quantification in fluorescence microscopy studies

Research papers will, by necessity, include 'representative' images that are selected to illustrate a point. They should also give sufficient information to validate the image selection and interpretation, e.g. by reporting number of replicate experiments and number of cells examined. This might include more rigorous quantitative data, e.g. numbers, size and relative position of features, or counts of the proportion of cells, organelles, etc. displaying a described feature.

As described in the primer on experimental design (Primer 2), scoring should ideally be performed 'blind', with either the identity of samples concealed from the observer to avoid potential bias or random images acquired and analyzed. Increasingly, the latter approach is achievable through high content imaging approaches, where automated acquisition and analysis enables imaging of multi-well slides or plates. In the case of time-course data, it is also useful if representative data (e.g. traces, movies) are supported by thorough quantitative analysis of a number of replicate experiments. This can include quantification of speed and direction of movement of objects in trafficking experiments – data which is notoriously difficult to assess 'by eye'.

Most fluorescence microscopy data is, at best, semi-quantitative, as it is difficult or impossible to relate fluorescence intensity to absolute numbers of molecules. However, this is not a significant problem in most cases, as relative concentrations of fluorescent-labelled molecules are sufficient to interpret the data reliably. In a few studies, comparison of labelling with the measured intensity of objects with known numbers of fluorescent molecules (e.g. viral particles, fluorescent microspheres) has been used to provide a more reliable quantitative measure. The issue of precise quantification of ions has been addressed by a different approach, where fluorophores such as the Ca^{2+} indicator, Fura-2, facilitate quantitation by ratiometric measurement rather than absolute intensities. In these cases, the absolute number of fluorophore molecules is not required for accurate quantitation (see Primer 17).

8.3.7 Time-course considerations and the benefits of live cell imaging

If fluorescence imaging studies present data from single (or few) time points, differences in kinetics may easily be overlooked. For this reason, live cell imaging has distinct advantages for many types of study, as it allows dynamics to be examined in more detail. Of course, such data can also overlook events that occur over a shorter timescale than the imaging interval (temporal resolution). In the case of GFP trafficking studies, a compromise is often made in the choice of time-lapse intervals to avoid the possible photo-damage and photo-bleaching that are likely if images are acquired too frequently. In live cell imaging experiments, it is also important to consider the condition of the cells as well as the level of potentially damaging illumination they are subjected to. Ideally, care will be taken to maintain temperature, humidity and, in some cases, CO_2 levels during live cell imaging to ensure that cells remain 'healthy' and avoid possible artefacts caused by changing conditions during their time on the microscope stage.

Complementary techniques

These are dealt with in the following primers of this book:

- Primer 8: Electron microscopy (EM and TEM)
- Primer 13: Immunocytochemistry
- Primer 16: GFP
- Primer 17: Fluorescence measurements of ion concentration
- Primer 21: In-situ hybridisation
- Primer 32: Axonal transport tracing

Further reading and resources

www.invitrogen.com/site/us/en/home/support/Tutorials.html (commercial website providing helpful introduction to the principles of fluorescence, filters, etc).

Conrad, C., Wünsche, A., Tan, T.H., Bulkescher, J., Sieckmann, F., Verissimo, F., Edelstein, A., Walter, T., Liebel, U., Pepperkok, R. & Ellenberg, J. (2011) Micropilot: automation of fluorescence microscopy-based imaging for systems biology. *Nature Methods* **8**,

246–249. (Compelling demonstration of the potential for automation of advanced fluorescence microscopy applications).

Keller, P.J., Schmidt, A.D., Santella, A., Khairy, K., Bao, Z., Wittbrodt, J., & Stelzer, E.H. (2010) Fast, high-contrast imaging of animal development with scanned light sheet-based structured-illumination microscopy. *Nature Methods* **7**, 637–642 (Explanation and demonstration of cutting edge light sheet based fluorescence imaging).

Molecular Probes handbook 11th edition (web version: www.probes.com/handbook/) (Includes introduction to fluorescence microscopy and comprehensive discussion of fluorescent probes and their applications).

Rossner, M. & Yamada, K.M. (2008) What's in a picture? The temptation of image manipulation. *Journal of Cell Biology* **166**, 11–15 (Entertaining and accessible explanation of the need to avoid inappropriate image manipulation and why proper documentation of image processing is necessary).

Schermelleh, L., Heintzmann, R., & Leonhardt, H. (2010) A guide to super-resolution fluorescence microscopy. *Journal of Cell Biology* **190**, 165–175 (Overview of theoretical and practical aspects of super-resolution microscopy).

Stephens, D.J. & Allan, V.J. (2003). Light microscopy techniques for live cell imaging. *Science* **300**, 82–86 (Review of live cell imaging techniques and types of microscope).

Swedlow, J.R. Hu, K., Andrews, P.D., Roos, D.S., & Murray, J.M. (2002) Measuring tubulin content in *Toxoplasma gondii*: a comparison of laser-scanning confocal and wide-field fluorescence microscopy. *Proceedings of The National Academy Of Sciences Of The United States Of America* **99**: 2014–2019 (Clear demonstration and explanation of why wide-field microscopy with deconvolution or confocal microscopy preferred for imaging certain specimens).

9
Intracellular 'Sharp' Microelectrode Recording

Helena C Parkington and Harold A Coleman

Physiology, Monash University, Australia

Changes in membrane potential are typically harnessed to functionally important responses of cells, tissues and, ultimately, the entire organism. The first step in understanding something is to observe it, and only then can one ask the 'how' questions. Intracellular glass microelectrodes were developed in the 1950s and remain our best and most reliable method of observeing the electrical behaviour of complex tissues.

9.1   Basic 'how-to-do' and 'why-do' section

Intracellular microelectrodes are used to measure the potential difference across the plasma membrane in living cells. Trans-membrane voltage (membrane potential) regulates the activity of many ion channels (e.g. voltage gated ion channels, Na_V, Ca_V, K_V and others) and transporters (e.g. Na^+/Ca^{2+} exchanger), and also that of several membrane-bound enzymes involved in signal transduction (e.g. phospholipase C, adenylyl cyclase), including some G-protein coupled receptors.

9.1.1   Advantages of intracellular microelectrodes

1. One of the most important advantages is the ability to study the function of a cell while the cell remains in its normal relationship with neighbouring cells, extracellular matrix, etc. In some important cases, electrical coupling between cells may be of interest. For example, in small blood vessels, the inner lining (endothelium) is electrically coupled to the outer smooth muscle cells (Primer 10, figure 10.1). This relationship cannot be studied following disruption of this coupling.

Essential Guide to Reading Biomedical Papers: Recognising and Interpreting Best Practice, First Edition.
Edited by Phil Langton.
© 2013 by John Wiley & Sons, Ltd. Published 2013 by John Wiley & Sons, Ltd.

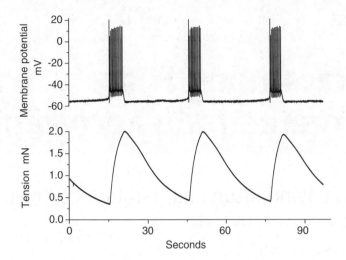

Figure 9.1 Simultaneous recording of membrane potential and tension from a strip of uterine smooth muscle. This segment was recorded 60 minutes after the initial impalement. The values of the resting potential and amplitudes of the spontaneously occurring action potentials indicate healthy cells, despite prolonged recording. Courtesy of Dr. Helena Parkington.

2. In many studies, cells are isolated from their surrounding tissue using enzymes, or enzymes are required to clean the surface of the cells to facilitate the Gigaohm seal required with patch-clamping. The effects of enzymes on properties of interest, usually involving transmembrane proteins with extracellular domains, are often poorly understood at best. Intracellular microelectrodes can be used to study such cells without having to use enzymes.

3. The fine tapering shaft of the microelectrode is likely to do minimal damage to the tissue through which it passes (but see below), and to flex with any movements (e.g. from movements associated with *in vivo* recordings), thus allowing for recordings from cells over prolonged times (hours – see Figures 9.1 and 9.2). The flexibility of the microelectrode also enables muscle contraction and membrane potential to be recorded simultaneously (Figure 9.1).

4. Unlike patch clamp (whole-cell configuration), sharp microelectrodes do not dialyse the cytoplasmic components (see Primer 11). Recordings with microelectrodes arguably inflict minimal, if any, disruption to intracellular messenger systems that may be critical for the regulation of ionic mechanisms in cells.

9.1.2   Disadvantages of intracellular microelectrodes

1. Impalement can cause stress on the cells, although this typically settles down over some minutes.

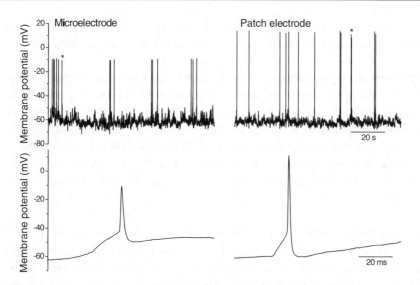

Figure 9.2   Recordings made in separate experiments from NPY neurons in the arcuate nucleus in slices of mouse brain. Both cells had input resistances of ≈1.4 GΩ, and input (membrane) time constants of ≈50 ms. Fluctuations in baseline are due to a high frequency of spontaneous synaptic activity that occasionally gave rise to action potentials. Recordings made with an intracellular microelectrode (left panels) had smaller action potentials due to electrode capacitance than those recorded with a patch electrode (right panels). Measures to reduce microelectrode capacitance were not vigorously pursued, since the amplitude of the action potential was not a critical parameter in the experiment. Action potentials marked by * in the upper panels are shown at an expanded scale in the lower panels. Activity recorded by the microelectrode remained reasonably constant for four hours, at which time the microelectrode was deliberately removed from the cell and the potential went to −6 mV. The membrane potential was corrected for this tip potential. The recording with the patch electrode was for about half an hour, by which time there was evidence of some run-down in activity. Courtesy of Dr. Helena Parkington.

2. The electrical properties of the microelectrode can cause some uncertainty in the absolute value of the membrane potential and can cause attenuation of some fast events (see below).

3. A very high concentration of potassium chloride in the microelectrode (e.g. 3 M) can result in excessive leakage into a small cell, so lower concentrations (e.g. 0.5 M KCl) should be used.

9.1.3   General characteristics of sharp intracellular microelectrodes

Most sharp intracellular microelectrodes are made by pulling a thin glass tube (capillary) to a very fine tip. The nature of the glass (borosilicate, aluminosilicate, quartz), the final diameter of the tip, the taper at the tip and the solution used to fill the electrode will all have major impacts on the characteristics of a microelectrode and on its suitability for membrane potential measurements in small cells.

Microelectrode tip and taper   For very small cells, a very small tip (typically 50–100 nm outside diameter) is needed if damage to the impaled cell is to be minimized. Fabrication of these electrodes requires a microelectrode puller that has a ring-shaped platinum heating element and electrical solenoids that can apply force to pull the pipette blank as the heating element softens the glass. The fabrication of microelectrodes requires skill and experience, and electrode quality is a significant factor in determining the quality of the recordings that can be achieved.

Electrode capacitance   Two conductors, separated by an insulator, will exhibit electrical capacitance. Glass is an excellent insulator and the salt solutions on either side of a microelectrode are conductors, so it can be readily demonstrated that microelectrodes behave as capacitors. Rapidly changing signals can pass through capacitors. This means that rapid events detected by microelectrodes can be shunted to earth across the wall of the microelectrode before they reach the amplifier circuitry in the headstage. Essentially, they are not seen or recorded. The main consequence of this is that very fast events, such as action potentials in neurons, are stunted in amplitude (see Figure 9.2), though the slower action potentials of smooth muscle cells (c.f. Figure 9.1) are unlikely to be affected. In most studies, the absolute level to which the action potential depolarizes is not critical, though large fast action potentials are very reassuring when considering the quality of an impalement.

Passing current   Microelectrode amplifiers incorporate the ability to pass current through the microelectrode while simultaneously recording the potential. The potential drop across the microelectrode resistance is balanced out using a Wheatstone Bridge-like circuit, leaving the response of the cell membrane and on and off transients. The amount of balancing typically gives a direct measure of the microelectrode resistance and is the easiest and most accurate method of determining this parameter. This is used as an indication of the relative tip size of the microelectrode and as an indication that the electrode has broken or become blocked. If not correctly balanced, or imbalances are not taken into account, the result can be inaccurate determinations of the parameters, such as cell input resistance and depolarization required to reach the action potential threshold.

9.2   Required controls

Any experiment that relies on the skill of the experimenter will require controls and criteria to ensure that recordings that are accepted as 'reliable data' conform to a standard that can be explained and defended.

1. The article's methods section should specify the criteria for accepting a microelectrode recording. If no such criteria are given, you should assume that minimal quality control has been applied; therefore, be cautious and be sceptical. Recordings from cells that have obvious and characteristic electrical events (e.g. action potentials) are simple to validate, but electrically quiescent tissues (e.g. vascular smooth muscle) present a greater challenge and more rigorous criteria for acceptance.

2. The article should directly show or cite earlier work that demonstrates that the preparation behaves 'normally' over the period of time that experiments are performed. This may be with reference to other *in vitro* experimental methods – measurement of force, for example.

3. A microelectrode is filled with a solution of high ionic strength (usually 0.5–3 M KCl) to reduce the resistance. As the filling solution is *not* similar to the bath or to the cytoplasm, the electrochemical gradient will support a net flux of ions. This flux gives rise to the development of a voltage (liquid junction potential) across the tip of the electrode. The size of the potential that develops can be as large as 15 mV and is complicated by other factors (Schanne *et al.*, 1968). The article should directly measure the tip potential or should cite earlier work and confirm that the conditions are not changed.

4. The flux of potassium ions from the tip of a sharp microelectrode, filled with 3 M KCl, has been estimated to be about 10^{-15} moles/s. In a very small cell, this could result in a significant change in cytoplasmic concentrations of both potassium and chloride. For this reason, it is preferable to use lower concentrations (e.g. 0.5 M KCl). The anion with K^+ can be critical in some situations. This particularly applies to cells in which the normal physiology relies upon a low intracellular chloride activity. In this case, the authors should have replaced the chloride with another anion (such as acetate, citrate or sulphate) to prevent changes in intracellular chloride concentration. It would be prudent for the authors to look for and so rule out time-dependent changes that could result from increasing intracellular potassium or chloride concentration.

9.3   Problems and pitfalls in interpretation and execution

From the foregoing discussion, several issues require careful scrutiny when absorbing information on intracellular recordings from the literature:

1. Recordings of biological electrical signals require that wires are connected to what are essentially salt solutions. This is generally accomplished using a

silver wire that has a chloride coat. If these electrodes are not carefully maintained, the voltage signal can drift (i.e. alter) over quite short periods of time. This drift makes unequivocal measurements of membrane potentials quite problematical.

2. Breaking into the cytoplasm, whether with a sharp microelectrode or a patch electrode, inflicts some damage to the cell, and recovery requires a longer time when using sharp microelectrodes. Thus, data from impalements at less than about 15 minutes must be viewed with some caution.

3. Values of resting membrane potential in absolute terms may need to be interpreted with caution when large tip potentials are apparent. It is good practice to find that studies use the average of the potential differences recorded upon impalement and withdrawal of the electrode. Certainly, authors should not ignore the potential recorded after withdrawal as it may be very far from zero. How this is handled may be included in the criteria mentioned in 'Required controls' above.

4. The amplitude of rapid changes in membrane potential, such as neuronal action potentials, may not be faithfully recorded. One of the principal causes of the attenuation that is observed is electrode capacitance. Changes in amplitude should not be assumed to be real without ruling out the effects of electrode capacitance.

5. Noisy traces must be viewed with suspicion. Fluctuations in membrane potential occur, but these must be on a 'clean' trace, with a good signal-to-noise ratio. The inherent noise in a well-designed and well-maintained microelectrode setup should be less than 0.5 mV.

6. Although critical cytoplasmic components are unlikely to disappear up into a microelectrode, perturbations in K^+ and Cl^- levels can occur over time. The authors should at least discuss this possibility and use logic or experimental controls to discount the impact on the interpretation of the data (see 'Required controls' above).

7. Sometimes electrode movement can produce artefactual changes in the membrane potential trace.

8. The small amplitude of biological potentials (mV) versus the hundreds of volts powering laboratory equipment (including, amplifiers, solution pumps, stim-ulators, lighting, etc.) means that special care must be taken to reduce interference. It is now quite uncommon to find published recordings that have significant non-biological noise. The overall level of noise in the recordings should be less than 1 mV, although cells with high input resistance can have quite 'noisy' membrane potentials. The great benefit of optimizing the equipment for low noise is the ability to observe small events.

9.3.1   Alternatives to sharp microelectrodes

1. **Potential sensitive dyes:** The loading of cells with potential-sensitive dyes can permit recording of membrane potential changes *in vivo* and in large contracting tissues, such as heart (Efimov *et al.*, 2010).

   Advantages and disadvantages of optical methods over sharp micro-electrodes: Whereas optical methods can provide information regarding the spread of membrane potential change within complex tissues, problems with interpretation require considerable skill and experience in the experimenters (Efimov *et al.*, 2010).

2. **Patch electrodes:** This technique utilizes glass electrodes with tips of much larger diameter and have a steep taper. A Gigaohm seal forms between the electrode tip and the plasma membrane, in contrast to the insertion of the tip of a sharp microelectrode through the plasma membrane into the cell (see Primer 11).

   Advantages and disadvantages of patch electrodes over sharp micro-electrodes: The larger tip size of the Gigaseal electrode permits greater current flow and, hence, greater control in voltage-clamp experiments. However, in whole-cell mode, cytoplasmic constituents can be lost into the very much larger volume of the pipette unless steps are taken (see Primer 11). To facilitate Gigaseal formation, the plasma membrane of the cell may need to be cleaned, necessitating isolation of the cell from its tissue. This places constraints on the questions that can be addressed. However, cells within tissues may be studied using Gigaseal electrodes in some tissues (e.g. brain slices) to considerable effect.

9.3.2   Comparison between sharp microelectrode versus patch electrode recordings

In a detailed and elegant study, Li *et al.* (2004) made a direct comparison between sharp microelectrode and patch electrode recordings (in current clamp mode) in spinal neurons ($\approx$20 µm in diameter) in frog tadpoles. The results showed essentially few differences between the two techniques. Two differences were apparent:

- First, the action potentials recorded with sharp microelectrodes were attenuated, compared with action potentials recorded via the patch electrode (see our recordings in Figure 9.2). This was most likely due to the greater capacitance of the sharp intracellular microelectrodes (see above).

- Secondly, there appeared to be some evidence of [1]damage (i.e. injury) upon microelectrode penetration, which largely, but not totally, recovered over the

---

[1] Damage tends to result in depolarization of the membrane potential.

following minutes. It is possible that, with recording for longer than the 3–15 minutes in this study, the recovery may have been more complete.

9.4   Complementary and/or adjunct techniques:

- Primer 10: Single electrode voltage-clamp (SEVC)

- Primer 11: Patch-clamp

- Two electrode voltage-clamp for large cells (e.g. oocytes).

Further reading and resources

Attwell, D. & Cohen, I. (1978). The voltage clamp of multicellular preparations. *Progress In Biophysics and Molecular Biology* **31**, 201–245.

Brennecke, R. & Lindemann, B. (1974a). Theory of a membrane-voltage clamp with discontinuous feedback through a pulsed current clamp. *Review of Scientific Instruments* **45**, 184–188.

Brennecke, R. & Lindemann, B. (1974b). Design of a fast voltage clamp for biological membranes, using discontinuous feedback. *Review of Scientific Instruments* **45**, 656–61.

Brown, K.T. & Flaming, D.G. (1977). New microelectrode techniques for intracellular work in small cells. *Neuroscience* **2**, 813–827.

Coleman, H.A., Tare, M. & Parkington, H.C. (2001). K^+ currents underlying the action of endothelium-derived hyperpolarizing factor in guinea-pig, rat and human blood vessels. *Journal of Physiology* **531**, 359–373.

Efimov, I.R., Fedorov, V.V., Joung, B. & Lin, S.F. (2010). Mapping cardiac pacemaker circuits: methodological puzzles of the sinoatrial node optical mapping. *Circulation Research* **106**, 255–71.

Finkel, A.S. & Redman, S. (1984). Theory and operation of a single microelectrode voltage clamp. *Journal of Neuroscience Methods* **11**, 101–127.

Li, W.C., Soffe, S.R. & Roberts, A. (2004). A direct comparison of whole cell patch and sharp electrodes by simultaneous recording from single spinal neurons in frog tadpoles. *Journal of Neurophysiology* **92**, 380–6.

Ramón, F., Anderson, N., Joyner, R.W. & Moore, J.W. (1975). Axon voltage-clamp simulations. A multicellular preparation. *Biophysical Journal* **15**, 55–69.

Richter, D.W., Pierrefiche, O., Lalley, P.M. & Polder, H.R. (1996). Voltage-clamp analysis of neurons within deep layers of the brain. *Journal of Neuroscience Methods* **67**, 121–131.

Schanne, O.F., Lavallee, M., Laprade, R. & Gagne, S. (1968). Electrical properties of glass microelectrodes. *Proceedings of the IEEE* **56**, 1072–1082.

10
Single electrode voltage-clamp (SEVC)

Harold A Coleman and Helena C Parkington

Physiology, Monash University, Australia

10.1 Basic 'how-to-do' and 'why-do' section

The great power of the voltage-clamp technique in elucidating ionic mechanisms, particularly voltage- and time-dependence, was well demonstrated by the landmark studies of Cole, and Hodgkin & Huxley and colleagues in the 1940s and 1950s. During those and subsequent studies on animal and plant cells over the following decades, the voltage-clamp was applied using two electrodes, one measuring membrane potential while the other passed current across the membrane. The requirement to insert two electrodes into a cell limited the use of this technique to relatively large cells.

The eventual development of the patch-clamp technique in the early 1980s overcame a number of the difficulties, enabling many types of cells, over a large range of size, to be voltage-clamped with a single electrode. The relatively large tip size ($\approx 1\,\mu m$ diameter) and the resulting very low access resistance (typically $\approx 2-5\,M\Omega$), compared with the input resistances of cells ($100-1,000\,M\Omega$), meant that a single electrode could be used both to measure membrane potential and also to pass current.

Prior to the development of the patch-clamp technique, a different approach, based on a *switching amplifier*, was developed by Brennecke & Lindemann (1974a, 1974b) to enable voltage-clamping of small cells with a single microelectrode. This approach is variously referred to as single-electrode voltage-clamp (SEVC), switching voltage-clamp, or single microelectrode voltage-clamp.

Essential Guide to Reading Biomedical Papers: Recognising and Interpreting Best Practice, First Edition. Edited by Phil Langton.
© 2013 by John Wiley & Sons, Ltd. Published 2013 by John Wiley & Sons, Ltd.

10.1.1   Advantages of SEVC

The SEVC technique combines the advantages of microelectrodes, listed in Primer 9, with the power of the voltage-clamp technique to study ionic mechanisms in small cells or in electrically short syncytial tissues. It can also enable the voltage-clamping of cells that cannot be accessed with patch-electrodes, such as neurons within deep layers of the brain (Richter *et al.*, 1996). The flexibility of the micro-electrode also enables contractile tissues, such as smooth muscle, to be voltage-clamped despite contraction movements of the cells (Figure 10.1).

In other situations, there can be too much connective tissue to enable patch-clamp recordings from the cells, and/or it is undesirable to use enzymes to clean the cells. In these cases, SEVC can come to the rescue. Importantly, SEVC recordings cause minimal, if any, disruption to intracellular messenger systems that may be critical for the regulation of ionic mechanisms in cells.

10.1.2   Limitations of SEVC

The high resistance of microelectrodes and the discontinuous, switching nature of SEVC limits the amount of current that can be passed by the electrode, thereby preventing the clamping of large currents. Furthermore, the capacitance of the microelectrode (see Primer 9) limits the speed of the SEVC, making it difficult to impossible to clamp very fast events.

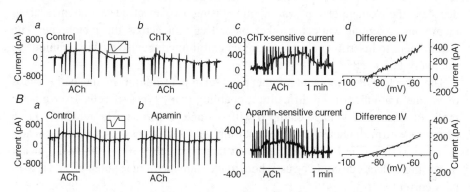

Figure 10.1 Membrane currents in electrically short segments of guinea-pig submucosal arterioles recorded under voltage-clamp with single intracellular microelectrodes (SEVC). Acetylcholine activated receptors on the endothelium to evoke endothelium-derived hyperpolarizing factor (EDHF). The underlying current involves intermediate- and small-conductance Ca^{2+}-activated K^+ channels, blocked by charybdotoxin and apamin, respectively, and whose I/V relationships, obtained from periodic ramps, are well described by the Goldman-Hodgkin-Katz equation for a pure K^+ current. The results show that no other currents in the endothelium or smooth muscle cells contributed to the outward hyperpolarizing current. The recording of the EDHF current required that the endothelium be intact and electrically coupled to the outer, single layer of smooth muscle cells. This was only achievable by the use of microelectrode SEVC. Coleman HA, Tare M & Parkington HC. (2001). K^+ currents underlying the action of endothelium-derived hyperpolarizing factor in guinea-pig, rat and human blood vessels. *J. Physiol.* **531**: 359–373. © Physiological Society.

10.1.3   Circumventing the limitations

Voltage-clamping of all kinds generally involves various compromises, and SEVC is no exception. The essential point is to focus on the critical physiology of interest. For example, the very wide range of potentials, over which the membrane is often varied with the patch-clamp technique (e.g. -100 to $+100$ mV), usually results in currents that are so large at the extreme potentials that it is difficult to discern what happens at the physiologically relevant potentials (which lie between -80 and 0 mV for many cells). SEVC can often provide very important information over this physiologically relevant range of potentials. SEVC control can be improved by the judicious use of ion channel blockers, and ion substitution can be used to reduce the concentrations of ions carrying currents that are not the focus of study.

10.1.4   Applying the SEVC technique

The theory and operation of SEVC is well described by Finkel & Redman (1984). However, a brief explanation is given here to enable a better understanding of its limitations and possible pitfalls.

During SEVC, the switching amplifier alternately '*samples*' the membrane potential and then passes current that is in proportion to the difference between the command potential and the sampled potential. The frequency of switching is manually adjustable, but current is passed during about one-third of the total cycle time. The current passes through both the microelectrode and the cell membrane, resulting in potential differences from these two sources. Once the current passing stops, the potential decays according to the membrane time constant and the microelectrode time constant.

A major requirement of SEVC is that the potential on the microelectrode decays much faster than that on the cell membrane. Thus, by the end of the 70 per cent non-current passing period, the electrode potential needs to have decayed to essentially zero, while there should only be a small change in the potential of the cell membrane. At this time, the potential is sampled and compared with the command potential in order to determine the amplitude of the next current passing phase. The rate of switching has to be carefully adjusted to ensure that the potential on the microelectrode has sufficient time in which to decay totally before the next cycle starts. The quicker the potential decays off the microelectrode, the greater the switching frequency that can be used, and therefore the faster the clamp response.

10.2   Pitfalls

The greatest problem with SEVC is that it is very easy to voltage-clamp the microelectrode but *not* the cell or preparation. Although published reports should not contain data that is so empty of biological significance, you need to be on your guard. You need to seek evidence that the authors recognize this risk and that they took appropriate steps to assure themselves (and you) that the data in their report were reliably real.

Failure to clamp more than the electrode can result from a failure to recognize a necessary compromise between the speed of the clamp (maxim = 'switching faster is better') and the requirement for the electrical potential difference of the electrode to decay fully before the value of remaining potential (i.e. potential of the impaled cell) is sampled and the switch is made to current passing (maxim = 'switching *too* fast is bad').

In practice, the experimenter should monitor the voltage at the headstage of the amplifier, using an oscilloscope at a fast sweep speed in order to observe the characteristic shape of the decay of the potential from the microelectrode, and adjust the switching frequency to ensure that the potential really does decay to zero. In any event, the article should describe how the switching frequency was optimized so that it was fast, but *not* too fast!

The crucial factors in determining the switching rate and quality of SEVC are the electrical properties of the microelectrode, and authors need to state how these were optimized. The use of microelectrodes with as low a resistance and as steep a taper as possible for a particular tissue will reduce both electrode resistance and capacitance. Electrode capacitance can be further reduced by minimizing the depth of physiological solution in the bath, and by treating the outside electrode surface with a hydrophobic compound.

There is a further complication. The risk of the problem described above is minimal if the electrode resistance is very low compared with the input resistance of the cells under study. However, it is not uncommon to have electrode resistances and cell/tissue input resistances that are comparable in magnitude and, in this situation, the issue becomes critical. Journal articles will include in the data the values for electrode resistance and estimated input resistance of the cell or preparation, but they will not put them conveniently together unless they believe it important to do so.

10.2.1   Indications of poor clamping

There are several signs that indicate that the clamping is poor, which may result from a switching frequency that is too high. In the more extreme case, the current trace will be the inverse of the expected membrane potential response. For example, depolarization may result in the current trace appearing like an upside-down action potential. In the case of voltage-activated sodium and calcium channels, a good indicator is the negative slope region of current-voltage (I/V) relationships (Figure 10.2).

As the membrane is stepped to increasingly more depolarized potentials, there should be a graded increase in the amplitude of these currents, as shown in Figure 10.2. Under conditions of poor clamp control, there can be a sudden, all-or-nothing-like increase in the amplitude of the inward current, and this will show up as a very steep negative slope in the I/V relationship. It can also occur if the

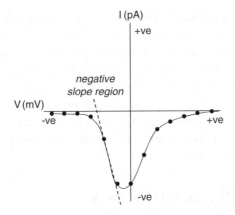

Figure 10.2   Current-voltage (I/V) relation typical of voltage gated calcium current. The dashed line indicates the part of the current voltage relation that is referred to in the text as a 'region of negative slope'. This simply means that, as the voltage becomes more positive, the conductance becomes increasing negative (i.e. the slope of the relation of current as a function of voltage is negative). Courtesy of Dr. Helena Parkington.

electrode cannot pass enough current to clamp the membrane, and/or the cell/tissue is not sufficiently isopotential. That is, active responses may occur in electrically distant cell processes that are too far from the electrode to be clamped.

The time course of tail currents can also indicate problems with poor clamp control. Usually, tail currents decay with an exponential time course. However, in the presence of a large series resistance, the current can almost 'linger' at a large amplitude before declining to zero in a non-exponential-like manner. These issues also apply to all voltage-clamp techniques, including patch-clamp, and are discussed in detail in a number of papers dating back to the earlier days of voltage-clamping (Attwell & Cohen, 1978; Ramón *et al.*, 1975).

In summary, SEVC has some significant advantages. So long as it is applied within its limitations, and particular care is taken with adjusting the clamp controls (such as clamp gain, negative capacitance, phase, filtering and switching frequency), SEVC can provide important information on ionic mechanisms in small cells and/or tissues that are not amenable to other forms of voltage-clamping. Thus, SEVC should be seen as a complementary voltage-clamp technique that adds to an electrophysiologist's armamentarium.

10.3   Alternative techniques
10.3.1   Gigaseal electrodes

This technique utilizes glass electrodes with tips of much larger diameter and having a steep taper. A Gigaohm seal forms between the electrode tip and the plasma membrane, in contrast with the insertion of the tip of a sharp microelectrode through the plasma membrane into the cell (see Primer 11).

Advantages and disadvantages of Gigaseal over sharp microelectrodes The larger tip size of the Gigaseal electrode permits greater current flow and, hence, greater control in voltage-clamp experiments. However, in whole-cell mode, cytoplasmic constituents can be lost into the very much larger volume of the pipette unless steps are taken (see Primer 11). To facilitate Gigaseal formation, the plasma membrane of the cell may need to be cleaned, necessitating isolation of the cell from its tissue. This places constraints on the questions that can be addressed, but cells within tissues may be studied using Gigaseal electrodes in some tissues (e.g. brain slices) to considerable effect.

10.3.2   Potential sensitive dyes

The loading of cells with potential-sensitive dyes can permit recording of membrane potential changes *in vivo* and in large contracting tissues, such as heart (Efimov *et al.*, 2010).

Advantages and disadvantages of optical methods over sharp microelectrodes Whereas optical methods can provide information regarding the spread of membrane potential change within complex tissues, problems with interpretation require considerable skill and experience in the experimenters (Efimov *et al.*, 2010).

10.4   Comparison between sharp microelectrode versus patch electrode recordings

In a detailed and elegant study, Li *et al.* (2004) made a direct comparison between sharp microelectrode and patch electrode recordings (in current clamp mode) in spinal neurons ($\approx 20\,\mu$m in diameter) in frog tadpoles.

The results showed essentially few differences between the two techniques. However, two differences were apparent:

- First, the action potentials recorded with sharp microelectrodes were attenuated, compared with action potentials recorded via the patch electrode (see our recordings in Figure 10.3). This was most likely due to the greater capacitance of the sharp intracellular microelectrodes (see above).

- Second, there appeared to be some evidence of [1]damage (i.e. injury) upon microelectrode penetration, which largely, but not totally, recovered over the following minutes. It is possible that, with recording for longer than the 3–15 minutes in this study, the recovery may have been more complete.

---

[1] Damage tends to result in depolarization of the membrane potential.

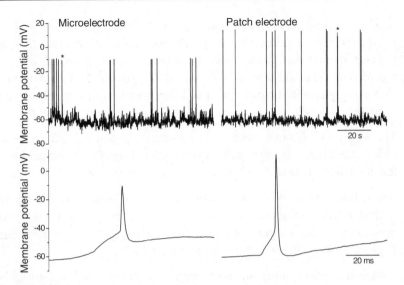

Figure 10.3 Recordings made in separate experiments from NPY neurons in the arcuate nucleus in slices of mouse brain. Both cells had input resistances of ≈1.4 GΩ, and input (membrane) time constants of ≈50 ms. Fluctuations in baseline are due to a high frequency of spontaneous synaptic activity that occasionally gave rise to action potentials. Recordings made with an intracellular microelectrode (left panels) had smaller action potentials due to electrode capacitance than those recorded with a patch electrode (right panels). Measures to reduce microelectrode capacitance were not vigorously pursued, since the amplitude of the action potential was not a critical parameter in the experiment. Action potentials marked by ∗ in the upper panels are shown at an expanded scale in the lower panels. Activity recorded by the microelectrode remained reasonably constant for four hours, at which time the microelectrode was deliberately removed from the cell and the potential went to −6 mV. The membrane potential was corrected for this tip potential. The recording with the patch electrode was for about half an hour, by which time there was evidence of some run-down in activity. Courtesy of Dr. Helena Parkington.

Additionally in this study, the microelectrode was buzzed in, and it is likely that gentle tapping to make the impalement would have produced less damage. The tip potentials in this study, −14 mV, are relatively large, perhaps due to the nature of the preparation. In many preparations, removal of the microelectrode from the cell under study results in the potential returning to within a few mV of zero potential.

10.5 Issues in the literature

From the foregoing discussion, several issues require careful scrutiny when absorbing information on intracellular recordings from the literature.

1. Breaking into the cytoplasm, whether with a sharp microelectrode or a patch electrode, inflicts some damage to the cell, and recovery requires a longer time when using sharp microelectrodes. Thus, data from impalements at less than about 15 minutes must be viewed with some caution.

2. Values of resting membrane potential in absolute terms may need to be interpreted with caution when large tip potentials are apparent. It is good practice to find that studies use the average of the potential differences recorded upon impalement and withdrawal of the electrode. Certainly, authors should not ignore the potential recorded after withdrawal, as it may be very far from zero.

3. The amplitude of rapid changes in membrane potential, such as neuronal action potentials, may not be faithfully recorded. One of the principal causes of the attenuation that is observed is electrode capacitance.

4. Noisy traces must be viewed with suspicion. Fluctuations in membrane potential occur, but these must be on a 'clean' trace, with a good signal-to-noise ratio. The inherent noise in a well-designed and well-maintained microelectrode setup is not much more than 1 mV.

5. Although critical cytoplasmic components are unlikely to disappear up into a microelectrode, perturbations in K^+ and Cl^- levels can occur over time. The authors should at least discuss this possibility and should use logic or experimental controls to discount the impact on the interpretation of the data.

6. Sometimes electrode movement can produce artefactual changes in the membrane potential trace.

10.6   Complementary and/or adjunct techniques

- Primer 9: Microelectrode recording.

- Primer 11: Patch-clamp in either voltage- or current-clamp modes.

- Two electrode voltage-clamp for large cells (e.g. oocytes).

Further reading and resources

Attwell, D. & Cohen, I. (1978). The voltage clamp of multicellular preparations. *Progress In Biophysics and Molecular Biology* **31**, 201–245.

Brennecke, R. & Lindemann, B. (1974a). Theory of a membrane-voltage clamp with discontinuous feedback through a pulsed current clamp. *Review of Scientific Instruments* **45**, 184–188.

Brennecke, R. & Lindemann, B. (1974b). Design of a fast voltage clamp for biological membranes, using discontinuous feedback. *Review of Scientific Instruments* **45**, 656–61.

Brown, K.T. & Flaming, D.G. (1977). New microelectrode techniques for intracellular work in small cells. *Neuroscience* **2**, 813–827.

Coleman, H.A., Tare, M. & Parkington, H.C. (2001). K^+ currents underlying the action of endothelium-derived hyperpolarizing factor in guinea-pig, rat and human blood vessels. *Journal of Physiology* **531**, 359–373.

Efimov, I.R., Fedorov, V.V., Joung, B. & Lin, S.F. (2010). Mapping cardiac pacemaker circuits: methodological puzzles of the sinoatrial node optical mapping. *Circulation Research* **106**, 255–71.

Finkel, A.S. & Redman, S. (1984). Theory and operation of a single microelectrode voltage clamp. *Journal of Neuroscience Methods* **11**, 101–127.

Li, W.C., Soffe, S.R. & Roberts, A. (2004). A direct comparison of whole cell patch and sharp electrodes by simultaneous recording from single spinal neurons in frog tadpoles. *Journal of Neurophysiology* **92**, 380–6.

Ramón, F., Anderson, N., Joyner, R.W. & Moore, J.W. (1975). Axon voltage-clamp simulations. A multicellular preparation. *Biophysical Journal* **15**, 55–69.

Richter, D.W., Pierrefiche, O., Lalley, P.M. & Polder, H.R. (1996). Voltage-clamp analysis of neurons within deep layers of the brain. *Journal of Neuroscience Methods* **67**, 121–131.

Schanne, O.F., Lavallee, M., Laprade, R. & Gagne, S. (1968). Electrical properties of glass microelectrodes. *Proceedings of the IEEE* **56**, 1072–1082.

11
Patch Clamp Recording

Neil Bannister and Phil Langton
Physiology & Pharmacology, University of Bristol, UK

11.1   Basic 'how-to-do' and 'why-do' section

Patch clamp recording includes a range of distinct recording techniques or configurations which enable the experimenter to investigate the electrophysiological responses of either intact cells (whole-cell recording configuration) or measure ion movements through single channels in isolated cell membrane patches (patch configurations). Patch clamp recordings can be performed on isolated cells or from tissue preparations (*in vitro* or *in vivo*); see Neher & Sakmann (1992) for a general overview.

Most cells have a small resting voltage across their plasma membrane in the order of 50–100 mV ($1\,mV = 10^{-3}\,V$). This membrane potential comes about as a result of a differential distribution of ions across the membrane and a selective permeability of the membrane to certain ions. Fluctuations in membrane potential come about due to the opening and closing of specific ion channels, embedded in the plasma membrane of the cell, activated by a range of stimuli. Patch clamp recording techniques enables the experimenter to monitor the activity of these ion channels either directly (by recording currents passing through ion channels) or indirectly (by recording changes in the membrane potential brought about by currents passing through them).

11.1.1   Patch pipettes

Recording electrodes (usually called patch pipettes) differ from those used for sharp electrode recording (see Primers 9 and 10) in that they have much larger tips ($\approx 1\,\mu m$). This is because they are designed to seal onto a small area of cell membrane rather than pierce it, as in sharp electrode recording. They are fabricated in a similar way to sharp microelectrodes.

Essential Guide to Reading Biomedical Papers: Recognising and Interpreting Best Practice, First Edition.
Edited by Phil Langton.
© 2013 by John Wiley & Sons, Ltd. Published 2013 by John Wiley & Sons, Ltd.

Critical to the technique is the formation of a tight (or 'sticky') seal between the tip of the pipette and the cell membrane. Selecting patch pipettes of the correct tip diameter and making sure the tip of the pipette remains clean are important determinants in obtaining tight seals. Many experimenters estimate the tip diameter of the pipette from its electrical resistance by applying a pulse of known current or voltage and calculating its resistance using Ohm's Law (R = V/I). Some researchers *fire polish* the tip of the pipette to aid seal formation.

11.2   Patch recording configurations

The first stage is to obtain a tight electrical seal between the tip of the patch pipette and the cell membrane. The formation of the seal is usually established by applying small suction. The tightness of the seal is monitored by applying a small pulse of electrical current down the pipette and measuring the voltage drop across the pipette. Electrical resistance can thus be calculated using Ohm's law (R = V/I). Seal resistances in the $G\Omega$ (10^9 Ohms) range (referred to as a *Gigaohm seal*) are necessary for patch clamp recordings. There are four principal recording configurations (Figure 11.1):

- Whole-cell

- Cell attached

- Inside-out

- Outside-out

11.2.1   Recording from the whole cell

This *cell-attached* configuration (used in its own right to record currents through single channels – see later) is the first step for establishing any of the other recording configurations. To enable recordings of the entire cell (*whole-cell configuration*), the membrane at the tip of the pipette must either be physically ruptured or made permeable by the introduction of ion channel-forming anti-biotics into the patch pipette in a variant known as '*perforated patch*' whole-cell (see Horn and Marty, 1988).

The transition between cell-attached and whole-cell can be monitored by applying a brief voltage command pulse. This is because the amount of current needed to charge the capacitance of the whole cell is much larger than that to charge the patch of membrane underlying the patch pipette. When this patch of membrane is ruptured to enter whole-cell mode, a characteristic whole-cell capacitance transient is observed. In this whole-cell recording configuration, the pipette solution is in direct contact (*ruptured patch*) or electrical contact (*perforated patch*) with the interior of the cell.

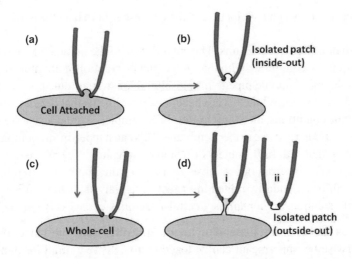

Figure 11.1 Cartoon representation of the four principal configurations of the patch clamp technique. (a) Cell attached; all patch clamp configurations start with this configuration.
(b) Isolated (inside-out) patch. This is achieved by moving the pipette away from the cell. In this configuration, the bath solution becomes the *de facto* intracellular solution. From the membrane patch's perspective, the inside surface of the membrane is facing out, hence 'inside-out'.
(c) Whole-cell. This is achieved from the cell attached configuration by the application of either negative pressure or a large electrical pulse (zapping or buzzing). An alternative to the 'conventional' whole-cell configuration, the 'perforated' patch whole-cell configuration is achieved by placing a pore-forming antibiotic into the pipette solution that will increase the conductance of the membrane and so permit the entire cell to be voltage-clamped.
(d) Isolated (outside-out) patch. This is achieved from the conventional whole-cell configuration. After achieving whole-cell configuration, the electrode is 'slowly' withdrawn. The next step is a bit of magic that is easy in some cells and next to impossible for others – the membrane is deformed into a tube that breaks and reseals to form an isolated patch. In this case, however, the outer surface of the membrane faces the bathing solution, hence 'outside-out'. Courtesy of Dr. Neil Bannister and Dr. Phil Langton.

Whole-cell recordings can be made in one of two modes:

- Current clamp

- Voltage-clamp

In *current clamp* mode, changes in membrane voltage are recorded. This often represents the physiological response of the cell to changes in ion channel activity. Because of the design of the recording amplifier, it draws negligible current from the recorded cell and therefore behaves as a near-perfect voltmeter.

In *voltage-clamp* mode, the membrane voltage is 'clamped' to a user-defined voltage (*the command voltage*). Specialized feedback circuitry in the recording amplifier passes current through the recording electrode to maintain the membrane potential at the command voltage. This holding current is recorded by the amplifier. Changes in ion channel activity are measured as changes in current required to 'clamp' the membrane voltage at the command voltage.

Direct access between the pipette and the cell brings both benefits and limitations:

1. **Control of ionic conditions.** The composition of the intracellular environment will quickly equilibrate with that of the pipette solution. Care must be taken in formulating a suitable pipette filling solution of similar ionic content to that of the recorded cell. While this is a benefit in experiments that depend critically on the ionic conditions, there are caveats. For example, calcium levels are known to vary, particularly close to the membrane. This can happen in any compartment of the cell that exhibits limited diffusion – so-called 'fuzzy areas'. Potential limitations arise from our relative ignorance of the importance of other factors that will be essentially diluted in the cell (see dialysis, below). It is typical for pipette solutions to contain key cellular factors such as ATP and GTP.

2. **Access that does not depend on membrane diffusion.** Substances that would otherwise be impermeant can be loaded into cells (e.g. lipid insoluble drugs, buffers for pH and calcium, peptides, antibodies, fluorescent dyes etc.).

3. **Dilution, dialysis or washout.** Although a patch pipette might contain only a few hundred microlitres, the cell volume is to the pipette volume as the drop is to the bucket. Anything inside the cell that can diffuse freely will be diluted and will disappear from the cell at a rate that depends on its mass (larger equals slower) and its affinity for fixed structures in the cell. This diffusion loss or dilution is often termed *dialysis*[1], and it can lead to a gradual loss (*run-down*) or, less often, gradual gain (*run-up*) of ion channel activity as mobile substances important for the regulation of activity are diluted by the large volume of pipette solution. This is sometimes referred to as 'washout'.

11.2.2   Recording from patches of cell membrane

These techniques allow the study of *single channels*. There are three isolated patch configurations:

- Cell-attached

- Outside-out (excised, or 'isolated')

- Inside-out (excised, or 'isolated')

Data from single channel recordings can be analyzed and presented in many ways – too many to list here. This primer is intended to give very basic information about

---

[1] Dialysis of cells may cause problems (e.g. loss of receptor function, or channel run-down), in which case a variant of the whole-cell configuration, known as 'perforated' patch, can be used. In this variant, the membrane patch remains physically intact and electrical access is achieved by the inclusion in the patch pipette solution of substances (often recognizable as antibiotics) that cause the membrane patch to become leaky, e.g. amphotericin B or nystatin or alpha toxin (see Horn & Marty, 1988).

the conformations of patch clamp and to alert you to some issues to bear in mind during your reading.

The currents recorded through single channels are extremely small (a few picoamperes, $pA = 10^{-12} A$). For this reason, the recording noise is the most significant limiting factor on one's ability to observe these brief, small amplitude events – essentially, the signal-to-noise ratio. Sources of noise include not only the electrical circuitry of the recording equipment but also the resistance of the seal between the pipette and the cell and the capacitance of the pipette itself. It can generally be assumed that the smaller the ionic currents that are of interest, the more effort has been expended on reducing all of these sources of noise.

Cell-attached mode   The membrane patch remains intact in cell-attached mode, so represents the least invasive mode as the contents of the cell's cytoplasm are not disturbed. However, there is evidence that some second messenger signalling pathways are not able to modulate the activity of channels inside the patch of membrane beneath the pipette. This does suggest some level of disruption. The cell-attached mode is relatively easy to obtain and tends to be stable over time. However, the environment on either side of the patch cannot be easily manipulated over the course of an experiment, and the membrane potential of the cell is intact and unknown.

Excised patches   Excised patches provide ultimate control for the experimenter. They are formed, perhaps surprisingly, by drawing the electrode away from the cell after cell-attached mode or whole-cell modes have been established. A potential disadvantage of excised patch configurations is that channels (AND receptors) have been isolated from the internal contents of the cell and this can interfere with the proper regulation of ion channel activity. If we assume that the pipette is called 'in' and the recording chamber is called 'out' and that cytoplasmic side on the membrane is called 'inside' and the extracellular side of the membrane is called 'outside', then:

- Outside-out patch – *outside* of the membrane faces *out* into the chamber

- Inside-out patch – the *inside* of the membrane faces *out* into the chamber

11.3   Required controls

Some may appear obvious; if so, then you are learning.

- Patch clamp is a powerful technique but it is hugely reductionist, particularly when considering isolated patch recordings, in which one is potentially studying a single molecule. Studies that utilize patch clamp should not make grand claims of functional significance without the corroboration from other experimental approaches that are more integrative (i.e. more functional). This is less a control than it is a notion of overall experimental design.

- It is common to use conditions that isolate the ionic current(s) of interest. If drugs are used to inhibit ion channels that are not the focus of study, care should be taken to ensure that these drugs do not alter the behaviour of the ion channels that *are* the focus of study. It is too often assumed that drugs have a greater specificity than it is safe to assume. One approach is to use two drugs that have selectivity for a given target. If the results of these experiments are equivalent, then it is less likely that they have unintended effects on the ionic current of interest. If these drugs come from different chemical classes, this control is more robust.

11.4   Pitfalls in execution or interpretation
11.4.1   Whole-cell recordings

Be aware that recordings made using patch pipettes clearly disturb the intracellular properties of the recorded cell. One should always consider this in interpreting research findings. A number of experimental phenomena (e.g. induction of long-term potentiation or LTP in brain slices) are known to be sensitive to washout of intracellular components. Alternative intracellular recording techniques could be employed to investigate this, such as perforated patch recordings or sharp electrode recording (see Primer 10).

11.4.2   Recording in voltage-clamp mode

There are a number of sources of error when using voltage-clamp to record membrane currents.

- **Space clamp errors.** Perfect voltage-clamp of the whole cell is only possible in very small 'electrically compact' (i.e. spherical) cells. Regardless, voltage-clamp recordings are often made from cells that are more complex, such as neurons that have an extensive dendritic arbor (e.g. pyramidal cells of the cortex and hippocampus). In such cases, the currents arising from ion flow through membrane ion channels (e.g. post-synaptic receptors) will not be optimally clamped, so the currents measured at the recording site (usually the cell body) will be distorted (slowed and attenuated) and thus will not accurately represent the true current profile. Depending on the experimental question, this may be an factor that should be considered by the authors and by you.

- **Electrode resistance.** The electrical resistance of the recording pipette (referred to as '[2]series resistance' or 'access resistance', R_a) can have several

---

[2] Although the terms *'series resistance'* and *'access resistance'* are often used interchangeably, they are different. Access resistance is actually a component of the series resistance but, in most cases, the access resistance is by far the largest component, so there is often very little difference between them.

effects on the interpretation of currents recorded under voltage-clamp conditions:

1. The first problem is *voltage error*. This occurs as a result of the relation between current, resistance and voltage that was referred to above as Ohm's Law. Increasingly, large currents passed through the '*access resistance*' of the pipette generates an increasing voltage that is termed a '*voltage drop*'. As a result of this, the membrane potential of the cell (the true potential) can differ from the command potential applied to the pipette. Under most recording conditions, this is usually small. However, when recording large ionic currents, or if R_a is very high, it can be large, leading to significant errors in assessing the true relation between membrane potential and membrane current. If R_a is known, the voltage error (difference between the cell voltage and the pipette voltage) will be directly proportional to the membrane current amplitude.

2. The second problem is the filtering effect of R_a combined with the capacitance of the cell membrane. The cut-off frequency of this low-pass filter can be very low (100s of Hz). It is clearly not reasonable to investigate rapid membrane events if the time-course of the membrane currents is being filtered heavily. In general, the higher the resistance, the slower and smaller the recorded current becomes. It is important to monitor the access resistance during an experiment, especially if one is monitoring the amplitude or kinetics of a membrane current over time. These issues really *should* have been addressed by a peer-review process of any rigor.

- **Temperature.** Most patch clamp experiments are performed at room temperature. Occasionally this needs to be taken into consideration by the reader if the data are being used to place limits on function. Cell signalling and ion channel activity can alter in very bizarre ways when temperature is altered (see Pabbathi *et al.*, 2004).

11.4.3   Recording from isolated patches of membrane

The principal problems that you may find in the literature can be summarized as:

Sampling errors   These are difficult, if not impossible, to exclude. You might consider these examples;

- *Ion channel distribution* – tends not to be uniform, so patches of membrane should *not* be regarded as representative (i.e. 'fair') samples of the population of ion channels in the membrane of that cell.

- *Ion channel activity* (spatial considerations) – may not be entirely uniform; a subset of channels may behave differently. This may be evident in whole-cell recordings, but the single channel correlate can evade recording.

- *Ion channel activity* (temporal considerations – e.g. modal gating) – patch-clamp recordings typically last only a few minutes. One has to assume that activity recorded during this recording phase is representative (or normal of activity over a longer time period). This is not always a safe assumption. If patches contain more than one channel (and most patches do), there is no opportunity to gather data that would validate this assumption. There are studies that suggest that the activity of single ion channels can vary in a cyclical manner, with some cycles of activity lasting several hours (see McManus and Magleby, 1988).

Recording (amplifier) errors   Amplifiers that are used to record patch clamp data are very sensitive, *but*:

- *Temporal resolution.* Patch recordings cannot faithfully record channel currents that are very brief. Good papers will mention this, will give a value for the 'minimum open time' resolution and will estimate the proportion of events that have been excluded (i.e. lost) from their analysis. One must remember that such short duration openings may be important functionally, and their exclusion is an analytical convenience to allow quantification and statistical comparisons.

- *Amplitude resolution.* It should be self-evident that channel currents need to be large enough to see and identify (resolve). Not all channels generate currents that are large enough for individual channel openings to be identified, although their presence can be inferred. Such currents can be studied using the whole-cell configuration, but little can be done using single channel configurations[3].

Logical errors   If evidence of channel regulation is shown only for outward current at $+100$ mV, it is *not* safe to assume that similar regulatory changes occur at more physiologically relevant membrane potentials – especially if the channel current is [4]*reversed* under such conditions.

11.5   Complementary and/or adjunct techniques

- Primer 9: Intracellular microelectrode recording.

- Primer 10: Single electrode voltage-clamp.

Further reading and resources

The Axon Guide for Electrophysiology & Biophysics – a practical laboratory guide.

---

[3] Macro patches are an exception to this. Macro patches are excised patches that incorporate large amounts of membrane, so that the activity of many hundreds or thousands of channels is recorded.

[4] *Reversed* in the sense that the flux of ions is reversed (i.e. opposite) from that expected under 'normal' conditions.

Available for download at various internet sites. Currently available for download at:

- www.whitney.ufl.edu/BucherLab/techspecs/Axon_Guide.pdf

- www.culturacientifica.org/textosudc/Axon_Guide.pdf

Chad, J. & Wheal, H. (Eds) (1991). Cellular Neurobiology: A Practical Approach. Oxford University Press.

Horn, R. & Marty, A. (1988). Muscarinic activation of ionic currents measured by a new whole-cell recording method. Journal of General Physiology 92, 145–159.

Neher, E. & Sakmann, B. (1992). The patch clamp technique. Scientific American **266**(3), 44–51.

McManus, O.B. and Magleby, K.L. (1988). Kinetic states and modes of single large-conductance calcium-activated potassium channels in cultured rat skeletal muscle. The Journal of Physiology, 402, 79–120.

Pabbathi, V.K., Suleiman, M.S. & Hancox, J.C. (2004). Paradoxical effects of insulin on cardiac L-type calcium current and on contraction at physiological temperature. Diabetologia **47**(4), 748–52.

Wallis D.I. (Ed) (1992). Electrophysiology: A Practical Approach. Oxford University Press.

12
Production of Antibodies

Elek Molnár

Physiology & Pharmacology, University of Bristol, UK

12.1   Basic 'how-to-do' and 'why-do' section

Antibodies (also known as immunoglobulins (abbreviated 'Ig'); see Figure 12.1) are host proteins produced in response to the presence of foreign molecules in the body. They are gamma globulins that are present in blood or other bodily fluids of vertebrates and used by the immune system to identify and neutralize foreign objects, such as bacteria and viruses. The highly specific interaction of an antibody with an antigen forms the basis of all immunochemical techniques. Antibodies can bind to a wide range of chemical structures (e.g. proteins, peptides, nucleic acids, carbohydrates, lipids, small chemical groups) and can discriminate among related compounds. Small molecules such as peptides may not be immunogenic, and will not usually produce antibodies if they are injected into an animal.

To make antibodies against small molecules, they must be coupled to a large protein to form a *hapten* (partial antigen)-carrier complex (Harlow & Lane, 1988, 1998). The region of an antigen that interacts with an antibody is defined as an *epitope*. An epitope is not an intrinsic property of any particular structure, as it is defined only by reference to the binding site of an antibody. Antibodies bind to complementary antigens by three-dimensional recognition. The antibody-antigen complex is stabilized by non-covalent bonds (e.g. van der Waals attraction, hydrogen bonds, salt bridges, hydrophobic interactions, electrostatic forces, ion pairs).

The binding of antibodies to antigens is reversible, and the interaction will conform to an equilibrium reaction. Because antibodies can recognize relatively small regions of antigens, occasionally they can find similar epitopes on other molecules. This forms the molecular basis for cross-reaction, which is a process that is exploited during the production of various antibodies (Harlow & Lane, 1988).

Essential Guide to Reading Biomedical Papers: Recognising and Interpreting Best Practice, First Edition.
Edited by Phil Langton.
© 2013 by John Wiley & Sons, Ltd. Published 2013 by John Wiley & Sons, Ltd.

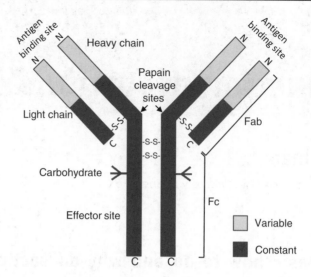

Figure 12.1 Schematic representation of antibody structure. Each antibody consists of four polypeptides: two large heavy chains and two small light chains connected by disulphide bonds (-S-S-) to form a Y-shaped molecule. While the general structure of antibodies is very similar, the amino acid sequence in the two upper ends of the Y varies greatly among different antibodies. These variable regions give the antibody its specificity for binding antigens. Therefore, each antibody molecule has two antigen binding sites. Digestion of the antibody with a protease (papain) yields three fragments, of which two (Fab) are identical and bind antigen, and a third (Fc), which comprises the effector domain. Antibodies are divided into five major classes (IgA, IgD, IgE, IgG and IgM), based on their constant region, structure and immune functions. Courtesy of Professor Elek Molnár.

When an antigen is introduced into a naïve animal, the first step in the generation of the primary response is phagocytosis of the antigen. Antigen-presenting cells degrade the antigen and display fragments of it on the cell surface. Helper cells associate with antigen-presenting cells by binding to the antigen fragment. Binding of a helper T cell to an antigen-presenting cell leads to proliferation of helper T cells. B cells also process antigens, but they do so in an antigen-specific manner. Binding of helper T cells to B cells is required for a strong antibody response and it also leads to proliferation of the B cells. Differentiation of B cells leads to the production of higher affinity antibodies. At the end of the primary response, the antigen is cleared, leaving primary memory cells. Subsequent injections of antigen induce a more potent response.

There are numerous factors that influence the strength and specificity of an antibody response:

1. Immunogenicity.

2. Preparation and presentation of the immunogen.

3. Techniques for injecting the immunogen to animals.

12.2   Primary antibodies

Primary antibodies may be polyclonal or monoclonal (Lipman *et al.*, 2005).

12.2.1   Polyclonal antibodies

The simplest method of producing primary antibodies against a foreign molecule is to immunize an animal (usually a rabbit, goat or guinea pig) against it. If it is a large molecule such as a protein, a peptide sequence from the protein is chosen for:

1. its accessibility (it is no use using a sequence that *in vivo* is inaccessible to the antibody due to its tertiary structure or its position in the hydrophilic (middle) region of a membrane); and

2. its unique sequence, as far as can be determined from sequence databases (sequences present in many different molecules are not good).

Repeatedly injecting the peptide, or other antigen, into the animal will trigger the immune system and produces antibodies against it. Blood serum which contains antibodies (usually IgG isotypes) is called *antiserum*. The animal may produce high concentrations of several antibodies against different (possibly overlapping) antigenic regions. Additionally, there will be very low concentrations of other antibodies that the animal may have been producing in the background, unrelated to the foreign antigen. The serum thus has high titres of a '*polyclonal primary antibody*', which is several antibodies directed against antigens.

The primary goal is to have a high antibody titre and high antibody affinity (binds strongly to antigens). IgG is the most desirable antibody, because of its favourable binding properties, its stability, its high concentration in serum and because it is simple to purify from antisera. First serum samples with antibodies are available 6–8 weeks after the start of the immunization.

Advantages   Relatively easy to make. Also, even if one of the epitopes re-occurs in a different molecule, the strength of staining for that molecule would be weaker than for the original injected antigen, due to the several different antibodies directed against different epitopes on the same antigen.

Limitations

1. It is not possible to know which epitope binds the antibody.

2. Due to individual variations of immune-responses induced in various host animals, the effectiveness and properties of polyclonal antibodies are highly variable, even for the same immunization protocol using the same antigen. Therefore, antibodies obtained from different host animals need to be

validated individually and used separately. This inherent variability of polyclonal antibodies is often responsible for fundamentally different results obtained with different batches of custom-made or commercially supplied antibodies.

12.2.2   Monoclonal antibodies

These are produced *in vitro* from single clones of (usually mouse) antibody-producing B-cells fused with myeloma cells. Each clone produces only one antibody (hence 'monoclonal'). An appropriate clone is selected, and this produces antibody against a single epitope. The fused cell is called a *hybridoma cell*, and it possesses the antibody-producing ability of the B-cell with the infinite life of the myeloma cell.

Advantages

1. High specificity for a single known epitope.

2. The hybridoma cells can be grown in a nutrient media and the antibodies can be harvested and purified from the supernatant.

3. Hybridoma cells can be frozen in liquid nitrogen and resurrected and grown when additional antibody is needed.

4. Monoclonal antibodies insure a consistent antibody with theoretically endless supply.

Limitations

1. The antibody may bind equally well to the epitope if it occurs in a different molecule.

2. Antibodies take 6–8 months to develop if the fusion of an appropriate antibody producing B-cell is successful.

12.3   Secondary antibodies

Secondary antibodies are antibodies used for the detection of the primary antibody in various immunochemical experiments (Primers 13, 14, 15, and see Figure 12.2). Secondary antibodies may be polyclonal or monoclonal, and they are available with specificity for whole Ig molecules or antibody fragments such as Fc or Fab regions. They are usually directed against the species-specific region on the antibody; hence, primary antibody produced in rabbit would require a secondary anti-rabbit antibody. The secondary antibody is usually conjugated to an enzyme (Figure 12.2), fluoro-chrome or colloidal gold particle to enable its visualization and detection (see Primers 7 and 8).

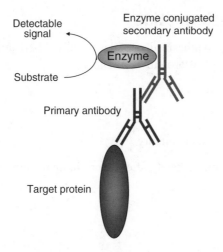

Figure 12.2   Schematic representation of primary and enzyme-conjugated secondary antibody reactions. After the primary antibody is bound to the target protein, a complex with enzyme-linked secondary antibody is formed. The enzyme (e.g. alkaline phosphatise, horseradish peroxidise), in the presence of a substrate, produces a detectable signal. This approach is commonly used in techniques such as ELISA, immunoblotting, immunocytochemistry and immunohistochemistry. Courtesy of Professor Elek Molnár. *A full colour version of this figure appears in the colour plate section.*

12.4   Choice of antigens for antibody preparation; their advantages and limitations

There are four main strategies:

1. Complete native protein

2. Synthetic peptide

3. Synthetic peptide using MAPS (multiple antigen presentation system)

4. Bacterially derived fusion protein

12.4.1   Complete native protein

Antigen   Purified native protein from tissue or transfected cells.

Advantages

1. Immunization with the native protein.

2. Can be done without knowing the amino acid sequence of the protein.

Limitations

1. Specific protein purification protocol is required.

2. Lack of subtype/subunit specificity.

12.4.2   Synthetic peptide

Antigen   Chemically synthesized peptide, usually 11–18 amino acids in length with a molecular weight of around 1–2 kDa.

Advantages

1. Great predetermined specificity.

2. Can be prepared immediately after determining the amino acid sequence of a protein.

Limitations

1. Short sequence reduces probability of reactivity with native protein.

2. Small size requires coupling to a carrier protein for antibody production.

3. The obtained antibodies must be affinity-purified.

12.4.3   Multiple antigen presentation system (MAPS)

This is used with synthetic peptides to improve antibody production (Molnar *et al.*, 1993; Fujita & Taguchi, 2011).

Antigens   MAPS peptides are synthesized as multiple copies on a branching lysyl matrix using solid-phase synthesis. Molecular weight of peptide complex is around 13–15 kDa for a 15 amino acid sequence.

Advantages

1. Preparation of the immunogen is quicker because no elaborate peptide purification is necessary.

2. No requirement for carrier protein.

3. Most of the immunogen consists of the amino acid sequence against which the antibody is to be raised.

4. Quantification of the immunizing dose is easier.

5. Higher anti-peptide antibody titres.

6. Much smaller quantity is required when compared with the unconjugated monomeric peptide for coating enzyme-linked immunosorbent assay (ELISA) plates.

Limitations

1. C-terminus of the peptide is attached to lysyl matrix and is not free, as in the native protein.

2. In some experiments, antisera to MAPs containing C-terminal peptides showed no cross-reactivity with the native protein.

3. Antibodies must be affinity-purified using monomeric peptide.

12.4.4   Bacterially derived fusion protein

Antigen   Long peptides representing significant portion of the native protein (see Figure 12.3; Pickard *et al.*, 2000).

Advantages

1. the large antigen has a great probability of adapting the conformation of the native protein.

2. Unlimited supply of antigen.

Limitations

1. Nucleotide sequence is required.

2. Lack of subtype/subunit specificity.

3. Technically demanding.

12.5   Purification of antibodies

The purity of the immunoglobulins is often critical, because other substances in the source material (e.g. antiserum, cell culture supernatant) may interfere with the detection process. For example, background labelling due to non-specific binding can interfere with the detection of the target antigen.

Antibodies can be purified by affinity chromatography or via precipitation. Antibodies are affinity purified with a protein A or protein G column that binds to the Fc portion of most immunoglobulins. Saturated ammonium sulphate precipitation is also frequently used to separate immunoglobulins (Harlow & Lane, 1988, 1998). Purification of specific antibodies can be achieved by immobilizing the antigen to a solid support surface in a way that the epitopes are available for antibody binding (Huse *et al.*, 2002; Figure 12.3). Following washing to remove non-specific proteins and contaminants, a low pH or high salt solution is used to

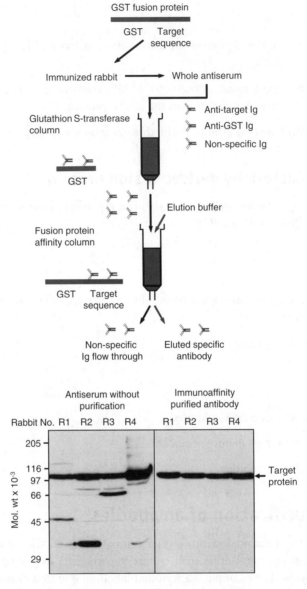

Figure 12.3 Purification of antibodies raised against a glutathione S-transferase (**GST**) fusion protein. The antiserum is first pre-adsorbed using an affinity column coupled with the GST protein. Any antibodies against GST are removed by this column. Specific antibodies to the target sequence flow into the second affinity column, which is coupled to a GST fusion protein containing amino acid sequences of the target epitope. Both columns allow the non-specific antibodies to flow through. Specific antibodies directed against the epitope region of the GST fusion protein are captured in the second column and eluted using a low pH buffer (0.1 M glycine-HCl, pH 2.5; red arrows). The elute is immediately neutralised by the addition of 1 M Tris-HCl (pH 8.0). Immunoblots illustrate reaction specificity of rabbit polyclonal antibodies before and after immunoaffinity chromatography purification. Full experimental details are available in Pickard *et al.* (2000). Courtesy of Professor Elek Molnár. *A full colour version of this figure appears in the colour plate section.*

transiently relax the interaction between the epitope and the antibody so that the antibodies can be recovered. The pH is then returned to neutral, and the purified antibody fraction is dialyzed into the appropriate buffer to regain its original conformation (Harlow & Lane, 1988, 1998; Figure 12.3).

12.6   Required controls

The development of immune response in the animal is analyzed in comparison with the serum obtained before the first inoculation with the antigen. It is essential to obtain pre-immune serum from each animal. The specificity of the primary antibody is usually established using immunochemical techniques (e.g. immunoblotting, ELISA, immunoprecipitation, etc.), immunocytochemistry and immunohistochemistry. It is important to prove that the obtained antibody cross-reacts readily and specifically with the native protein in the investigated tissue or protein extracts. Ideal control samples are:

1. transfected cells that express a high concentration of the investigated protein; and/or

2. tissue obtained from transgenic animals, which lack the target antigen.

Primary antibodies should be pre-absorbed with excess of the corresponding antigenic peptide or protein. Under these conditions, all immunoreactivity should disappear. While this control will show that the antibody binds to the antigen, it does not exclude the possibility that it does not bind to anything else.

12.7   Common problems and pitfalls in execution or interpretation

While there is a vast amount of literature available for each step of the antibody production process, and no shortage of advice for individual antibody production strategies, the final outcome is very uncertain. Most common problems include:

- Lack of immune response,

- While the antibody strongly interacts with the antigen used for the immunization, there is no cross-reactivity with the native protein.

- Non-specific cross-reactions with unrelated proteins.

- Limited supply of polyclonal antibodies.

- Individual variations of immunoresponse in different animals.

- High costs.

- Time-consuming.

The success rate can be improved somewhat by using:

- more than one antigen derived from the investigated target;
- different antigen presentation methods;
- different immunization protocols;
- larger numbers of animals;
- different species.

While huge number of antibodies are available commercially for a range of targets, and these are used extensively in various experimental approaches, investigators have to navigate through a complex system to find out which particular antibodies are appropriate for a specific experiment. At present, there is no universal validation for antibodies (Bordeaux *et al.*, 2010). An antibody that works in one system may perform poorly in other (see Primers 13 & 15 for more details). Therefore, researchers need to know exactly how these antibodies have been produced and characterized to determine whether an immunoreagent will work in the assay they are using.

Using the same antibody to investigate isoforms of target proteins in different species needs to be considered particularly carefully, due to potential inter-species differences in the corresponding epitope regions that may prevent specific immunochemical interactions. Therefore, if this information is available, sequences of the epitope regions need to be compared carefully before an antibody is used in a species different from the one it was originally developed for.

As discussed in greater details in Primer 13 (Immunocytochemistry and immunohistochemistry), antibodies display 'method specificity'. This means that some antibodies may work well in one protocol (e.g. immunoblotting – Primer 14) but fail to produce specific labelling in another (e.g. immunohistochemistry, Primer 13 or immunoprecipitation, Primer 14). This is due to the different presentation of epitopes under different experimental conditions. For example, proteins are denatured and unfolded for immunoblotting and epitopes are fully accessible for immunochemical interactions.

In contrast, fully folded proteins are used for most immunoprecipitation and immunolocalization experiments, where epitope regions may not be readily accessible on the surface of the target protein and antibodies are unable to bind to these regions. Fixation of samples for immunocytochemistry and immunohistochemistry experiments can significantly alter the structure of the epitopes, which can interfere with antibody binding (see Primer 13 for details).

There is a growing need for high-quality antibodies and more robust validation data. Furthermore, it is often disappointing that only few of the commercially available antibodies seem to work in independently performed tests. Without appropriate validation, misinterpretation of non-specific immunoreactions can lead to erroneous conclusions. The necessity of more rigorous specificity tests is becoming more widely accepted. Verification of immunoreaction specificity is particularly important in immunohistochemical experiments, where only few antibodies pass correctly applied specificity criteria (Blow, 2007; Pradidarcheep *et al.*, 2008; Michel *et al.*, 2009).

12.8   Complementary techniques

* Primer 13: Immunohistochemistry

* Primer 14: Immunoprecipitation

* Primer 15: Immunoblotting (Western blotting)

Cited work, further reading and resources

Blow, N. (2007). The generation game. *Nature* **447**, 741–744.

Bordeaux, J., Welsh, A.W., Agarwal, S., Killiam, E., Baquero, M.T., Hanna, J.A., Anagnostou, V.K. & Rimm, D.L. (2010). Antibody validation. *BioTechniques* **48**, 197–209.

Fujita, Y. & Taguchi, H. (2011). Current status of multiple antigen-presenting peptide vaccine systems: Application of organic and inorganic nanoparticles. *Chemistry Central Journal* **5**, 48.

Harlow, E. & Lane, D. (1988). *Antibodies: a laboratory manual.* Cold Spring Harbor Laboratory Press, Cold Spring Harbor, NY.

Harlow, E. & Lane, D. (1998). *Using antibodies: a laboratory manual.* Cold Spring Harbor Laboratory Press, Cold Spring Harbor, NY.

Huse, K., Böhme, H. & Scholz, G.H. (2002). Purification of antibodies by affinity chromatography. *Journal of Biochemical and Biophysical Methods* **51**, 217–231.

Lipman, N.S., Jackson, L.R., Trudel, L.J. & Weis-Garcia, F. (2005). Monoclonal versus polyclonal antibodies: distinguishing characteristics, applications, and information resources. *ILAR Journal* **46**, 258–268.

Michel, M.C., Wieland, T. & Tsujimoto, G. (2009). How reliable are G-protein-coupled receptor antibodies? *Naunyn-Schmiedebergs Archives of Pharmacology* **379**, 385–388.

Molnar, E., Baude, A., Richmond, S.A., Patel, P.B., Somogyi, P. & McIlhinney, R.A.J. (1993). Biochemical and immunocytochemical characterization of antipeptide antibodies to a cloned GluR1 glutamate receptor subunit: Cellular and subcellular distribution in the rat forebrain. *Neuroscience* **53**, 307–326.

Pickard, L., Noël, J., Henley, J.M., Collingridge, G.L. & Molnar, E. (2000). Developmental changes in synaptic AMPA and NMDA receptor distribution and AMPA receptor subunit composition in living hippocampal neurons. *Journal of Neuroscience* **20**, 7922–7931.

Pradidarcheep, W., Labruyère, W.T., Dabhoiwala, N.F. & Lamers, W.H. (2008). Lack of specificity of commercially available antisera: better specifications needed. *Journal of Histochemistry and Cytochemistry* **56**, 1099–1111.

13
Immunocytochemistry and Immunohistochemistry

Elek Molnár

Physiology & Pharmacology, University of Bristol, UK

13.1   Basic 'how-to-do' and 'why-do' section

See Primer 12 (Production of antibodies) for basic principles of immunochemical reactions.

In immunocytochemistry and immunohistochemistry techniques, an antibody is used to specifically label a cellular antigen (e.g. a protein). These localization approaches fundamentally rely on the high specificity, affinity and sensitivity of antibody-antigen interactions. Antibodies are visualized either directly or indirectly (usually via a secondary antibody), with a stain that is easily detectable under a light or electron microscope. An ideal stain is stable, will reveal accurately the distribution of the antigen and is suitable for both low- and high-resolution localization studies. By convention, the immunochemical detection of an antigen in tissues is known as immunohistochemistry, and detection in cells is termed immunocytochemistry.

13.2   Basic procedures

Immunohistochemical and immunocytochemical approaches involve complex, multi-stage procedures that show considerable variations in the literature. Here only the most commonly used general steps are outlined; see Burry (2010) for more detailed description of individual experimental procedures. Most frequently, the following key steps are used for immunolocalization.

First, tissue sections or individual cells are fixed to preserve morphology. Fixatives can damage antibodies, so they need to be removed and neutralized before immunolabelling. It is often necessary to permeabilize cell membranes (e.g. by

Essential Guide to Reading Biomedical Papers: Recognising and Interpreting Best Practice, First Edition.
Edited by Phil Langton.
© 2013 by John Wiley & Sons, Ltd. Published 2013 by John Wiley & Sons, Ltd.

using detergents or ethanol) to gain access to intracellular epitopes. To prevent non-specific binding of antibodies via random protein-protein interactions, samples need to be saturated ('blocked') with proteins (e.g. bovine serum albumin) before immunolabelling.

Fixed, permeabilized and blocked tissue sections or cells are then incubated with the primary antibody (usually immunoglobulin G – IgG), raised against the specific target antigen in one species (e.g. rabbit, guinea pig or mouse). Unless the primary antibody is labelled in some way to allow detection under the microscope (direct labelling), a second antibody, raised against IgGs of the first species (e.g. rabbit anti-mouse IgG), is then applied (indirect labelling).

After each step, samples are carefully washed to remove excess (unbound) antibodies. Various light and electron microscopic approaches are used for the visualization of different labels (described below).

13.3   General considerations

Optimal conditions for immunohistochemical or immunocytochemical detection must be determined for each individual situation (Pool & Buijs, 1988; Lorincz & Nusser, 2008). Antigen availability, antigen-antibody-complex affinity and stability and detection enhancement methods are often variable, with the specificity of the antibody very hard to prove, and there is currently no standardized method of quantification (Figure 13.1). The general concept of immunochemical reactions,

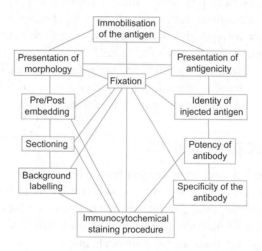

Figure 13.1   Mutual relationships between key factors influencing the outcome of an immuno-cytochemical staining procedure. Immunolabelling in each immunocytochemical procedure is the result of a complex set of interrelationships between antigen preparation, tissue processing and staining procedures. For example, extensive fixation for the maintenance of tissue morphology may interfere with the presentation of the target antigen via covalent modification of epitopes to such an extent that they are no longer recognized by the primary antibody. Figure based on Pool CW, Buijs RM (1988) Antigen identity in immunocytochemistry. In: Molecular Neuroanatomy (eds: van Leeuwen FW, Buijs RM, Pool CW and Pach O), Elsevier, Amsterdam, pp233–266. © Elsevier. Courtesy of Professor Elek Molnár.

antibody-antigen recognition, together with various antibody validation procedures and specificity controls, are described in Primer 12. In the following sections,
key issues are highlighted for each stage of the immunolocalization process.

13.3.1   Specimen preparation (fixatives, section type, antigen retrieval)

Fixatives are needed to preserve cell and tissue architecture in a reproducible and
lifelike manner. To achieve this, tissue blocks, tissue sections, cultured or acutely
dissociated cells or smears are immersed in a fixative fluid. In cases where whole
animal systems are studied, the animal is perfused with fixative via its circulatory
system, typically using 4 % paraformaldehyde. Fixatives stabilize cells and tissues,
thereby protecting them from the rigours of subsequent processing and staining
techniques.

Unfortunately, the methods that are best for the preservation of tissue/cell
structure do so by altering proteins. This could lead to masking of epitopes, where
the binding of antibody to target protein may be prevented (Figure 13.1). It is
important to try various fixatives and antigen retrieval methods. Fixatives may work
by several means: formation of cross-linkages (e.g. aldehydes such as glutaraldehyde or formalin) or protein denaturation (e.g. acetone and methanol) or a
combination of both. Fixation strengths and times must be optimized so that
antigens and cellular structures can be retained and epitope masking is minimal.

The next consideration for immunological staining is the type of section to use.
For immunohistochemistry, the common options are fixed or unfixed cryostat
(frozen) sections, fixed 'wet' or vibrotome sections, or fixed paraffin- (wax) or
resin-embedded sections. Fixed frozen sections are often quickest and easiest to use,
and they allow excellent antigen presentation. However, frozen sections give less
morphological detail and resolution. Fixed paraffin- or resin-embedded tissues have
better fidelity and clarity.

To facilitate the immunological reaction of antibodies with antigens in fixed
tissue, it may be necessary to unmask the antigens through pre-treatment of the
specimens. Antigen retrieval includes a variety of methods by which the availability
of the antigen for interaction with a specific antibody is maximized. Despite
appearing bizarre and destructive, the most common techniques are enzymatic
digestion or heat-induced epitope retrieval through microwave irradiation (Cuevas
et al., 1994), autoclaving or pressure cooking.

13.3.2   General considerations for antibody staining

Primary antibodies may be labelled with an enzyme (e.g. horseradish peroxidase or
alkaline phosphatase; Figure 13.2), fluorophore (e.g. FITC or rodamine; Figure 13.3)
or colloidal gold particles (Figure 13.4).

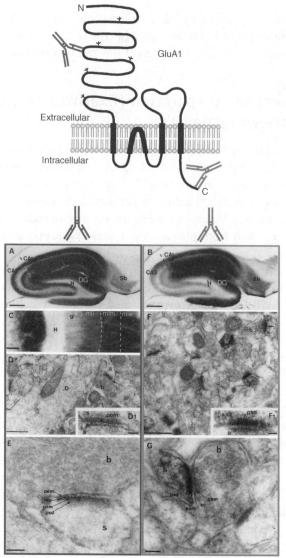

Figure 13.2   Correlated light and electron microscopy using site-directed anti-peptide primary antibodies in combination with horseradish peroxidise (HRP) enzyme-conjugated secondary antibody. Two rabbit polyclonal antibodies were generated against N-terminal (red) and C-terminal (blue) epitopes of the GluA1 AMPA receptor subunit protein using synthetic peptides (Molnar *et al.*, 1994). Immunolabelling with both antibodies indicates similar distribution for GluA1 in the rat hippocampus (A, B). Electron micrographs of GluA1 immunoreactivity (D-G) indicate that the HRP reaction end-product (rep) is concentrated at the extracellular face of the postsynaptic membrane (pom), demonstrating extracellular location for the N-terminal antibody binding site (E). In contrast, the reaction end-product obtained with the C-terminal antibody is located intracellularly at the post-synaptic membrane, on the postsynaptic density (psd), and not in the synaptic cleft (sc). Modified from Molnár *et al.* (1994). Scale bars: 0.5 mm (A and B), 50 μm (C), 0.5 μm (D and F), 0.05 μm (D1 and F1), 0.1 μm (E and G). Reproduced, with permission, from Molnár E, McIlhinney RAJ, Baude A, Nusser Z, Somogyi P (1994) Membrane topology of the GluR1 glutamate receptor subunit: Epitope mapping by site-directed anti-peptide antibodies. *J Neurochem* **63**:683–693. © John Wiley & Sons Ltd. Courtesy of Professor Elek Molnár. *A full colour version of this figure appears in the colour plate section.*

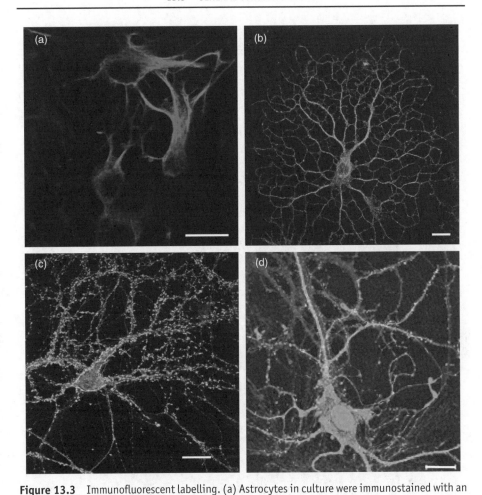

Figure 13.3 Immunofluorescent labelling. (a) Astrocytes in culture were immunostained with an antibody against glial fibrillary acidic protein (GFAP), which is a major component of intracellular skeleton of protein filaments in these cells (Luyt *et al.* 2003). PANEL (a) Reproduced, with permission, from Luyt K, Varadi A, Molnar E (2003) Functional metabotropic glutamate receptors are expressed in oligodendrocyte progenitor cells. J Neurochem **84**:1452–1464. (b) Differentiated rat oligodendrocyte in primary culture, stained with an antibody to metabotropic glutamate receptor mGluR5 (Luyt *et al.* 2006). PANEL (b) Reproduced, with permission, from Luyt K, Varadi A, Durant CF, Molnar E (2006) Oligodendroglial metabotropic glutamate receptors are developmentally regulated and involved in the prevention of apoptosis. J Neurochem **99**:641–656. (c) Co-localization of activated (autophosphorylated) calcium-calmodulin kinase II (CaMKII; green – an enzyme involved in long-term potentiation in the central nervous system synapses) and synaptotagmin (synaptic marker) in hippocampal neuronal cultures (Appleby *et al.* 2011). PANEL (c) Reproduced, with permission, from Appleby VJ, Corrêa SAL, Duckworth JK, Nash JE, Noël J, Fitzjohn SM, Collingridge GL, Molnár E (2011) LTP in hippocampal neurons is associated with a CaMKII-mediated increase in GluA1 surface expression. J Neurochem **116**:530–543. (d) Co-localization of flag epitope-tagged truncated C-terminal cargo domain of myosin VI (green), GluA1 (red) and GluA1-4 (blue) AMPA receptor subunits using mouse, rabbit and guinea pig primary antibodies, respectively (Nash *et al.* 2010). Scale bars: 10 μm. PANEL (d) Reproduced, with permission, from Nash JE, Appleby VJ, Corrêa SAL, Wu H, Fitzjohn SM, Garner CC, Collingridge GL, Molnár E (2010) Disruption of the interaction between myosin VI and SAP97 is associated with a reduction in the number of AMPARs at hippocampal synapses. J Neurochem **112**:677–690. © John Wiley & Sons Ltd. *A full colour version of this figure appears in the colour plate section.*

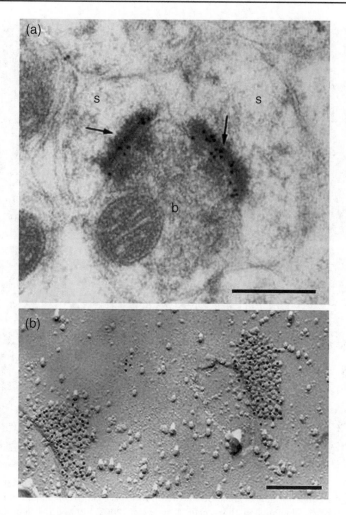

Figure 13.4 Localization of AMPA receptors using immunogold methods. (a) Post-embedding immunogold labelling of GluA1-4 AMPA receptor subunits in the hippocampus using a secondary antibody coupled to 10 nm gold particles. Modified from Nusser *et al*. (1998). PANEL A. Reproduced, with permission, from Nusser Z, Lujan R, Laube G, Roberts JDB, Molnar E, Somogyi P (1998) Cell type and pathway dependence of synaptic AMPA receptor number and variability in the hippocampus. *Neuron* **21**:545–559. (b) SDS-digested freeze-fracture replica labelling of AMPA receptors (10 nm gold particles) in the molecular layer of the cerebellum. Modified from Masugi-Tokita *et al*. (2007). Scale bars: 250 nm. PANEL B. Reproduced, with permission, from Molnár E (2011) Long-term potentiation in cultured hippocampal neurons. *Semin Cell Dev Biol* **22**:506–513. © Elsevier.

These labels allow the direct visualization of the antigen/antibody complexes using light and electron microscopic techniques. However, many primary antibodies are not labelled with some form of tags, and researchers need to rely on indirect visualization of unlabelled antibodies by enzyme, fluorophore or colloidal gold labelled secondary antibodies (see Primer 12 for details) or more complex indirect (e.g. biotin and streptavidin interaction-based) detection systems (Burry, 2010).

If a secondary antibody is used, it must be generated against the immunoglobulins of the primary antibody source (e.g. if the primary antibody is raised in rabbit, then the secondary antibody must be anti-rabbit). Using common secondary antibodies (e.g. anti-rabbit, anti-mouse, anti-guinea-pig, anti-goat, etc.) will reduce costs, as well as providing an opportunity for signal amplification if one primary antibody can be detected by multiple secondary antibodies or more complex indirect signal amplification systems (Burry, 2010). The optimal titre of both the primary and secondary antibody should be determined for each batch and for the system in which they are employed. The optimal antibody dilution will be that which gives the strongest specific antigen staining with the lowest non-specific background.

13.3.3 Antibody detection

The most commonly used detection methods are:

1. colorimetric (enzyme mediated);

2. fluorescence; and

3. colloidal gold (Burry, 2010; Masugi-Tokita & Shigemoto, 2007).

When choosing a substrate for enzyme mediated detection, one should select a compound that yields a precipitating product (e.g. diaminobenzidine, 4-chloro-1-naphthol or nitro blue tetrazolium). A molecule that fluoresces can be attached to the antibody for detection using excitation light. Examples are fluorescein, rhodamine, Texas Red, Cy3, Cy5 and Alexa Fluor (AF) dyes.

It is often necessary to perform dual or triple immune-labelling of the same sample to examine if two target proteins or a target protein and a marker protein of subpopulation of cells or cell organelles are co-localized. To avoid crossover of signals produced by two labels, both the excitation and emission wavelengths of simultaneously used fluorophores should be as far apart as possible in their light spectrum. This usually means that one secondary antibody is conjugated to a green fluorophore, while the other secondary antibody is labelled with a red fluorophore to offer a good separation of spectrum. Detection of antigens by antibodies conjugated to colloidal gold particles is often used for electron microscopy, and these may also be visualized at the light microscopic level (Masugi-Tokita & Shigemoto, 2007).

Signal amplification techniques greatly enhance the sensitivity of immuno-histochemical and immunocytochemical methods. The signal amplification methods (e.g. avidin-biotin system or poly-conjugated secondary antibodies) may be used in conjugation with either of the above detection techniques (Burry, 2010). When signal amplification is used to enhance the specific signal, however, one should be aware that non-specific signals (background labelling) may also become amplified.

Table 13.1 Summary of the advantages and limitations of various immuno-localisation approaches.

	Enzymatic methods	Fluorescent methods	Immunogold methods
Required	• specific primary antibody • horseradish peroxidase (or alkaline phosphatases) conjugated secondary antibody for indirect labelling	• specific primary antibody • fluorochrome-coupled secondary antibody for indirect labelling	• specific primary antibody • colloidal gold-coupled secondary antibody for indirect labelling • resin-embedded tissue (post-embedding reaction)
Advantages	• high sensitivity • enables correlated light and electron microscopy (Figure 13.2) • allows the identification of whether the epitope is intra- or extracellular, pre- or postsynaptic, etc. (Figure 13.2)	• high sensitivity • enables the co-localization of several antigens simultaneously (Figure 13.3c, d) • non-diffusible marker • allows quantitative comparison of different antigenic sites • enables the labelling of intact cells in culture (Molnár, 2011)	• high resolution ($\approx$20 nm) • quantifiable (particulate marker) • exposed tissue elements are equally accessible to antibodies • quantitative comparison between two antigenic sites is possible by using secondary antibodies coupled to different size colloidal gold particles
Limitations	• diffusible nature of the enzyme reaction end-product • often inappropriate for quantification (enzyme reaction is not linear)	• fluoroprobes can be analyzed only at the light microscopic level • low resolution ($>$0.2 um)	• low sensitivity • inappropriate for the identification of whether the epitope is intra- or extracellular

The advantages and limitations of various immuno-localization approaches, such as enzymatic (Figure 13.2), fluorescent (Figure 13.3) and immunogold (Figure 13.4) methods are summarized in Table 13.1.

The comparison of three key methods of immunogold electron microscopy, such as pre-embedding, post-embedding (Figure 13.4a) and sodium dodecyl sulphate (SDS)-digested freeze-fracture replica labelling methods (Figure 13.4b; Masugi-Tokita & Shigemoto, 2007) is presented in Table 13.2.

13.4   Required controls

Several controls are needed for the correct interpretation of immunolocalization experiments (Burry, 2000; Saper & Sawchenko, 2003; Holmseth *et al.*, 2006;

Table 13.2   Comparison of three key methods of immunogold electron microscopy (based on Masugi-Tokita & Shigemoto, 2007).

	Pre-embedding method	Post-embedding method (Figure 13.4a)	Sodium dodecyl sulphate (SDS)-digested freeze-fracture replica labelling (Figure 13.4b)
Advantages	• widely used, relatively straightforward • same specimen can be used for light and electron microscopy	• antigens exposed on the surface of the sections • immunoreactions occur evenly • detects antigens in less accessible cell compartments (e.g. neuronal synapses) • improved quantification	• high sensitivity • immunoreactions occur evenly on a two-dimensional replica membrane • epitopes are denatured by SDS • highly reproducible
Limitations	• antibodies do not penetrate evenly into tissue samples • antigens in less accessible cell compartments (e.g. neuronal synapses) are often undetectable • quantitative comparisons are difficult	• proteins are poorly accessible by antibodies • low sensitivity • difficult to discriminate the precise location of the antigen (e.g. extracellular/ intracellular location of plasma membrane proteins) • time-consuming reconstruction of serial ultrathin sections is required	• random fracturing • few morphological clues on the replica • separation of proteins to various surfaces in the replica ('P' or 'F' face; see Masugi-Tokita & Shigemoto, 2007) is unpredictable

Rhodes & Trimmer, 2006; Fritschy, 2008). First of all, it is essential to confirm the specificity of the primary antibody interaction with the target antigen using immunochemical techniques (immunoblotting, ELISA, immunoprecipitation, etc.) before it is used for immunohistochemical or immunocytochemical experiments (see Primer 12 for details). Omitting the primary antibody, or using sera obtained from the immunized animal before the first injection of the antigen ('pre immune serum') instead of the primary antibody, are frequently used negative controls to check the specificity of various immunolabelling procedures.

Furthermore, primary antibodies should be pre-absorbed with excess of the corresponding antigenic peptide or protein. Under these conditions, a large part of staining of the sections should disappear. While this control will show that the antibody binds to the antigen, it does not exclude the possibility that it may bind to something else. It is often very helpful to use two different antibodies raised against different epitopes of the same peptide or protein for localization studies (Figure 13.2). If patterns of staining are the same with both antibodies, this is strong circumstantial evidence in favour of specificity (Figure 13.2). However, it is often difficult to obtain two primary antibodies to the same target.

Data produced by complementary techniques frequently used for the validation of antibody specificity (e.g. immunoblotting, immunoprecipitation) can increase the likelihood of the correct target antigen being detected. The ideal control sample is tissue obtained from a transgenic animal which lacks the target antigen. Comparison of the distribution of immunoreactivity with images obtained by using other cellular imaging strategies (e.g. *in situ* hybridization or auto-radiography) could provide additional supporting evidence. For example, cells/regions/tissues where the presence or absence of the target antigen is expected based on these complementary imaging approaches can be used as positive and negative controls, respectively.

13.5   Common problems and pitfalls in execution or interpretation

It should *not* be assumed that commercially supplied antibodies are specific and properly tested! A well-conducted study will either refer to other studies that provide the necessary validation, or else will include the tests necessary to validate the antibody for their particular experimental needs. The importance of this is demonstrated neatly in the report of Lorincz & Nusser (2008).

The most common problems are:

- no staining of either controls or specimen;
- weak staining;
- high background staining.

Troubleshooting requires one to determine whether difficulties are related to specimen, antibodies, technique, environment or interpretation. This can be achieved by modifying the fixation conditions, antigen retrieval, blocking solution, antibody titrations and dilutions, washing of sections and sample and antibody storage conditions.

13.6   Complementary techniques

- Primer 12: Antibody production
- Primer 8: Fluorescence microscopy
- Primer 7: Electron microscopy
- Primer 21: *In situ* hybridization

Cited work, further reading and resources

Appleby, V.J., Corrêa, S.A.L., Duckworth, J.K., Nash, J.E., Noël, J., Fitzjohn, S.M., Collingridge, G.L. & Molnár, E. (2011). LTP in hippocampal neurons is associated with a CaMKII-mediated increase in GluA1 surface expression. *Journal of Neurochemistry* **116**, 530–543.

Burry, R.W. (2000). Specificity controls for immunocytochemical methods. *Journal of Histochemistry and Cytochemistry* **48**, 163–165.

Burry, R.W. (2010). *Immunocytochemistry: a practical guide for biomedical research.* Springer, New York.

Cuevas, E.C., Bateman, A.C., Wilkins, B.S., Johnson, P.A., Williams, J.H., Lee, A.H.S., Jones, D.B. & Wright, D.H. (1994). Microwave antigen retrieval in immunocytochemistry: a study of 80 antibodies. *Journal of Clinical Pathology* **47**, 448–452.

Fritschy, J. (2008). Is my antibody-staining specific? How to deal with pitfalls of immunohistochemistry. *European Journal of Neuroscience* **28**, 2365–2370.

Holmseth, S., Lehre, K.P. & Danbolt, N.C. (2006). Specificity controls for immunocytochemistry. *Anatomy and Embryology* **211**, 257–266.

Lorincz, A. & Nusser, Z. (2008). Specificity of immunoreactions: The importance of testing specificity in each methods. *Journal of Neuroscience* **28**, 9083–9086.

Luyt, K., Varadi, A. & Molnar, E. (2003). Functional metabotropic glutamate receptors are expressed in oligodendrocyte progenitor cells. *Journal of Neurochemistry* **84**, 1452–1464.

Luyt, K., Varadi, A., Durant, C.F. & Molnar, E. (2006). Oligodendroglial metabotropic glutamate receptors are developmentally regulated and involved in the prevention of apoptosis. *Journal of Neurochemistry* **99**, 641–656.

Masugi-Tokita, M. & Shigemoto, R. (2007). High-resolution quantitative visualization of glutamate and GABA receptors at central synapses. *Current Opinion In Neurobiology* **17**, 387–393.

Molnár, E., McIlhinney, R.A.J., Baude, A., Nusser, Z. & Somogyi, P. (1994). Membrane topology of the GluR1 glutamate receptor subunit: Epitope mapping by site-directed anti-peptide antibodies. *Journal of Neurochemistry* **63**, 683–693.

Molnár, E. (2011). Long-term potentiation in cultured hippocampal neurons. *Seminars in Cell and Developmental Biology* **22**, 506–513.

Nash, J.E., Appleby, V.J., Corrêa, S.A.L., Wu, H., Fitzjohn, S.M., Garner, C.C., Collingridge, G.L. & Molnár, E. (2010). Disruption of the interaction between myosin VI and SAP97 is associated with a reduction in the number of AMPARs at hippocampal synapses. *Journal of Neurochemistry* **112**, 677–690.

Nusser, Z., Lujan, R., Laube, G., Roberts, J.D.B., Molnar, E. & Somogyi, P. (1998). Cell type and pathway dependence of synaptic AMPA receptor number and variability in the hippocampus. *Neuron* **21**, 545–559.

Pool, C.W. & Buijs, R.M. (1988). Antigen identity in immunocytochemistry. In: van Leeuwen, F.W., Buijs, R.M., Pool, C.W. & Pach, O. (Eds) *Molecular Neuroanatomy*, pp 233–266. Elsevier, Amsterdam.

Rhodes, K.J. & Trimmer, J.S. (2006). Antibodies as valuable neuroscience research tools versus reagents of mass distraction. *Journal of Neuroscience* **26**, 8017–8020.

Saper, C.B. & Sawchenko, P.E. (2003). Magic peptides, magic antibodies: Guidelines for appropriate controls for immunohistochemistry. *Journal of Comparative Neurology* **465**, 161–163.

14
Immunoprecipitation (IP)

David Bates
Physiology & Pharmacology, University of Bristol, UK

14.1   Basic 'how-to-do' and 'why-do' section

Immunoprecipitation (IP) uses the properties of antibodies that bind to specific antigens to enable protein-protein interaction or post-translational modifications to be studied (see also Primers 12, 13 & 15). The principle of the technique is simple, in that it relies on the ability of antibody-antigen complexes to be precipitated from a solution, and the non-bound proteins washed off, before analysis by immunoblotting (IB – see Primer 15). The approach here will concentrate on two different uses:

1. Investigating protein complexes.

2. Post-translational modification.

14.1.1   Antibodies

Antibodies consist of a fixed chain and a variable chain. The fixed chain has a specific sequence for every antibody species and type; e.g. every mouse immunoglobulin G1, (Mouse IgG$_1$) has the same fixed chain (see Primer 12). The variable chain is specific for each antigen epitope, e.g. anti-VEGFR-2, anti-β-actin, etc. Antibodies can be added to samples, tissues, cells, homogenates, supernatants, etc. and will bind with very high affinity to their target epitope. Therefore, to immunoprecipitate a protein (e.g. the VEGF receptor), one needs to extract the protein from the cytoplasm, nucleus or membranes of cells, then add a sufficient amount of antibody to the homogenate so that the concentration will be high enough for most, if not all, of the receptor to be bound by antibody.

Essential Guide to Reading Biomedical Papers: Recognising and Interpreting Best Practice, First Edition. Edited by Phil Langton.
© 2013 by John Wiley & Sons, Ltd. Published 2013 by John Wiley & Sons, Ltd.

14.1.2   Protein extraction and cell lysing

Protein extraction for IP requires some considerations in addition to those for general protein extraction for immunoblotting (see Primers 13, 15 and 24). To extract proteins from cells, it is crucial to consider the location of the target protein within the cell.

Why? To extract membrane proteins (e.g. receptors, such as the VEGF receptor), for example, cells need to be treated with detergents (e.g. Triton or SDS) that effectively dissolve the lipid membrane. However, use of strong detergents, particularly under reducing conditions, can also start to break up protein complexes, particularly if they are dependent on membrane localization. For instance, protein kinase C (PKC) can be membrane associated due to attachment of lipid molecules to it during synthesis. If detergents are used, then the PKC may dissociate from membrane proteins that it may normally interact with, such as tyrosine kinase receptors. Therefore, the strength of the detergent used (concentration and type), as well as the presence of reducing agents (e.g. dithiothreotol), need to be optimized and tightly controlled as they have the potential to affect both results and interpretation.

To extract proteins from compartments within the cell (e.g. the nucleus), the cell must be broken open very carefully (e.g. by osmotic forces), and then the nuclear fraction separated by differential centrifugation. In this way, the proteins of interest can be enriched. This can be important because components of signalling complexes may differ by intracellular location (e.g. nuclear verses mitochondrial versus cell membrane). By way of example, the Notch receptor has an intracellular domain (ICD) which, when the membrane-bound receptor is activated, is cleaved and translocates to the nucleus. Once in the nucleus the ICD can activate transcription, and thus the types of the proteins binding to the ICD will depend on the cell fraction being examined.

Experiments that use total homogenate containing cell membrane, mitochondrial, nuclear and cytosolic targets would not be as accurate and informative as those using specific cellular fractions. It is therefore imperative that papers report adequate details of the methods of homogenation and fractionation. Reports should justify the strategies used by referencing earlier work (with at least a summary of the key factors), or else by demonstrating their optimization for the protein being investigated.

Understanding how proteins are regulated and how this regulation impacts on cell function is of fundamental significance for studies of both protein and cell function. It has been known for many years that cells contain kinases and phosphatases that catalyse, respectively, phosphorylation and dephosphorylation of target proteins. It is also very clear that phosphorylation of specific amino acid residues is a key method by which protein function is regulated. For this reason, tools that enable the study of phosphorylated proteins have been developed. It is now common for IP experiments to be conducted under conditions that prevent dephosphorylation (homogenation, etc. on ice, in the presence of phosphatase inhibitors such as vanadate). In some cases, use of freshly prepared (not frozen) cell lysate can benefit protein phosphorylation detection. This means that all samples should be prepared otherwise identically (including freeze/thaw cycles).

14.1.3   Recognition and precipitation

Once the tissue or sample containing the antigen of interest has been recognized by incubation with a suitable antibody, there are two ways of precipitating the complexes.

1. **Direct** (antibody-dependent). The sample can be centrifuged (agarose beads) or exposed to a magnet (paramagnetic beads), because the primary antibody used was bound to a bead that can be precipitated (e.g. agarose or paramagnetic bead). This is good when common antigens are used (e.g. antigens that define cell function, such as CD31 for endothelial cells), as the demand is such that these are commercially available pre-conjugated to the bead.

2. **Indirect** (protein A- or protein G-dependent). The sample is incubated with protein A or protein G that is covalently bound to paramagnetic or agarose beads. Protein A and protein G are bacterial proteins that bind with high affinity to specific regions of the fixed chain of antibodies, so will recognize all primary antibodies (see Primer 12). More recently, fusion proteins containing the antibody recognition regions from both proteins A and G have become available.

During incubation of the sample with the antibody bead or protein A/G bead, large insoluble complexes are formed that can either be spun to the bottom of the test tube using a centrifuge, or isolated by magnetic columns. This should be done at 4°C to prevent further interactions occurring.

14.1.4   Detection

The pellet that forms contains the antibody-antigen complex. This pellet is then washed with buffers and spun again. The procedure is usually repeated two to three times to remove most of the proteins that are not specifically bound by antibody and merely weakly associate with the complexes. These will generally be 'weakly associated' proteins, but this is not always the case. The [1]*stringency* of the washing conditions (buffers and temperature) can be controlled to ensure either that only tightly associated proteins complex, or more weakly associated proteins survive the washing protocols.

Let us assume that we now have a purified form of the target antigen, along with everything else that has bound to it and has survived the repeated washing. This

---

[1] **Stringency**: in this context it means 'effectiveness' or 'harshness'. A 'low stringency' wash will allow more weakly bound proteins to survive washing (i.e. remain bound) than will a 'high stringency' wash.

pellet is then denatured to break up the complexes, and run on an SDS-poly-acrylamide gel and subjected to IB (see Primer 15 for details). The choice of antibody for the immunoblot is dictated by the experimental aim – to investigate protein-protein interactions or post-translational modification.

Consider how this works by thinking through the two following examples:

Example 1. It is possible to use an antibody for IP that detects only modified amino-acids, such as a phosphotyrosine antibody or a phosphoserine antibody (antibodies that bind with high affinity *only* to phosphorylated tyrosine (pY) or phosphorylated serine (pS)). In the former case (pY), any protein that is detected on the blot is tyrosine phosphorylated and was precipitated by the phospho-tyrosine-specific antibody (anti-pY). Thus, if the precipitate is subjected to IB for VEGFR2, any signal must come from phosphorylated VEGFR2.

A word of caution: The approach above will also, however, detect on the IB membrane (see Primer 15) *any* proteins that were bound to VEGFR-2 during the IP and that have also been tyrosine phosphorylated. These will usually run at a different speed on the gel, so give rise to a band in a different position on the immunoblot (different apparent molecular mass). This is where the stringency of the wash after the IP stage becomes of obvious importance.

An alternative strategy is to use a VEGFR-2 antibody that does not discriminate phosphorylation state (a *global* or *total* antibody) to precipitate the VEGFR-2 protein and the use an anti-pY antibody on the immunoblot. In this way, it is possible to detect phosphorylated VEGFR-2.

Example 2. In this example, an antibody is used that detects an entirely different protein, to see if the second protein precipitates with the first. This is often referred to as co-immunoprecipitation (co-IP), or immuno-co-precipitation. It is used to test the notion that two or more proteins normally exist as a 'function complex' that involves some degree of physical association of the proteins. For instance, if one immunoprecipitates VEGFR-2 after activation and probe the blot with an anti-Shc antibody, one ought to obtain a band of the expected mass for Shc. The interpretation being that Shc appears to bind tightly to the VEGFR-2 receptor during activation.

14.2 Required controls

As with any technique, the number and relevance of controls determines the confidence of the interpretation. Appropriate controls include:

1. **Input control**. The strength of the band depends ultimately on the amount of protein available in the lysate, and this can vary between otherwise identical experiments. It follows that it is necessary to measure and control

for protein content. For example, if precipitating with a phospho-specific antibody and detecting a phosphorylated protein, it is necessary to control for differences in the protein content. It would be good practice for any lab that extracts protein from lysates to retain samples each lysate extraction to run on the final immunoblot, to ensure that there was no change in the amount of starting protein in the experimental and control lanes. (see Figure 14.2).

2. **Pre-clearing.** Non-specificity is a major problem in IP. To reduce non-specific binding of proteins to either protein A/G or to the beads, the lysate is often pre-treated with a mixture of protein A/G and agarose beads in the absence of any primary antibody. Proteins that bind (binding will be non-specific) to the beads are removed by centrifugation, leaving a now 'cleared' lysate that can be subjected to IP with the primary antibody. The degree of non-specificity (and the potential loss of antigen in pre-clearing) can be judged by recovering the pellet formed after pre-clearing and including a sample on the same immunoblot as the cleared samples.

3. **Non-specificity of the primary antibody** (negative control). As with all antibodies, controls for non-specific binding (see Primers 12 and 13) should be used to show that the proteins being precipitated reflect the specific interaction between primary antibody and antigen. Good controls for this include:

   - a non-specific immunoglobulin (e.g. mouse IgG);

   - a cell line/knockout that does not express the antigen as a negative control;

   - spiking a sample with recombinant protein;

   - a cell lysate in which the target is known to be expressed as a positive control.

For instance, if investigating VEGFR2 phosphorylation, human umbilical vein endothelial cells treated with and without VEGF (positive control) and HEK cells (negative control, as HEK cells do not express VEGFR2) are good positive and negative controls, but in addition, a mouse IgG primary with the HUVECs would be a good negative.

4. **Specificity of primary antibody.** To ensure that the IP has pulled down the specific protein of interest, the blot should be probed with an antibody to the same protein as the precipitating antibody. For example, if the IP is for VEGFR2 and the IB for pY, then the IB should also be done with a different anti-VEGFR2 antibody, ideally to a different epitope, to ensure specificity and effective pull-down.

14.3 Common problems and pitfalls in execution or interpretation

14.3.1 Specificity of effect

IP usually results in the precipitation of complexes of protein, not single proteins. Therefore, when the IP is with anti-VEGFR2 and the IB with anti-pY, as in the example above, a band should come up at around 180 kDa if the VEGFR2 is tyrosine phosphorylated. To have confidence that it is actually VEGFR2 that is being phosphorylated, the blot would subsequently (or coincidentally if using fluorescent IB) need to be probed with a (preferably different) antibody to VEGFR2 to show that it is the same band.

Alternatively, the IP stage can be performed with an anti-phosphotyrosine antibody (precipitating *all* proteins that have a phospho-tyrosine reside *and* are recognized by the antibody) and the blot probed with a VEGFR-2 antibody. Even at this stage, it is not possible to conclude that hormone A acted on receptor A to activate kinase A and so phosphorylate target A. For example, VEGF may have activated a different receptor that went on to phosphorylate VEGFR-2, or even inhibited a receptor that then prevented a phosphatase from acting on VEGFR-2. There are many possible routes between receptor and effector, and all must be considered by experiment.

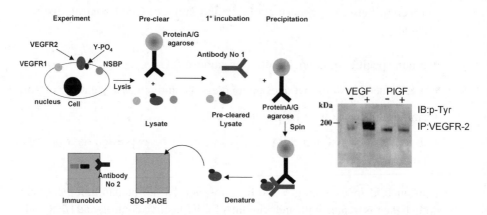

Figure 14.1 The steps involved in IP. A. After cell lysis, antigens that might bind non-specifically to protein A/G (non-specific binding protein NSBP), are removed by pre-clearing. Lysate is then incubated with primary (1°) antibody, the protein A/G agarose, spun, denatured and subjected to sodium dodecylsulphate-polyacrylamide gel electrophoresis (SDS-PAGE) and immunoblotting using a different primary antibody. B. An immuno- (Western) blot of a cell lysate of cells that have been exposed either to VEGF or PlGF, and then immunoprecipitated (IP) with an antibody to VEGFR-2. The protein has then been run on a polyacrylamide gel and immunoblotted (IB) with an antibody to phosphotyrosine (p-Tyr). It can be seen from the gel that VEGF, but not PlGF, can stimulate phosphorylation of one or two proteins around 200 kDa in these cells. VEGFR-2 is approximately 200 kDa, so this is interpreted as showing that VEGF, but not PlGF, can result in phosphorylation of VEGFR-2. Courtesy of Professor Dave Bates. *A full colour version of this figure appears in the colour plate section.*

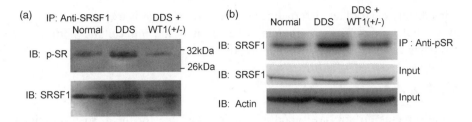

Figure 14.2   Illustration of a figure from an IP experiment. The splice factor SRSF1 is hyper-phosphorylated in podocytes from patients with a mutation in WT1 that results in Denys Drash Syndrome (DDS). (a) Protein was extracted from podocytes, immunoprecipitated (IP) with an SRSF1 antibody and immunoblotted (IB) with a pan-phospho SR antibody (top) or SRSF1 antibody (bottom). The band in the DDS cells is more intense in the p-SR blot than in the other cell types, indicating more phosphorylated SFRS1. The lower blot (SRSF1) shows that the amount of SRFS1 protein on the blot is the same. Note the lack of molecular weight markers sin the lower blot and lack of non-specific control IP. (b) Protein was extracted from podocytes and half immunopre-cipitated with an anti-phosphoSR protein antibody. Both the IP and the crude protein extract were immunoblotted with a SRSF1 antibody and Actin as a loading control. Again, the stronger band in the DDS indicates more phosphorylated SR protein; note lack of molecular weight markers and non-specific control antibody. Reproduced, with permission, from E M Amin, Oltean S, J Hua, MV Gammons, M Hamdollah Zadeh, GI Welsh, MK Cheung, L Ni, S. Kase, E.S. Rennel, KE Symonds, D. G. Nowak, B Pokora-Royer, M.A. Saleem, M. Hagiwara, V A Schumacher, S J Harper, DR. Hinton, D O Bates*, MR Ladomery*. Alternative VEGF splicing in WT1 mutants reveals SRPK1 to be a novel angiogenesis target. Cancer Cell 20(6):768–80 © Elsevier.

14.3.2   Duration and magnitude of response

IP gives a snapshot in time that equates to exactly when the protein was extracted. If there is a need to understand the time course of a response using IP, then a time sequence will need to be performed. This can provide important insight – for example, if the protein in question only exist in the phosphorylated state for a few minutes during a signalling event. This also underlines the enormous importance of controlling conditions of the IP experiments (time, temperature, cellular fractionation, etc.). For example, if the time between receptor stimulation and protein extraction is not strictly controlled for each replicate, the results of multiple replicates of what was intended to be a single time point may be either positive or negative. Considering the experiment in Figure 14.1, it may be that PlGF acts more rapidly than VEGF, so the receptor was phosphorylated at an earlier time point.

14.3.3   Protein extraction and modification

Methods for protein extraction vary and can alter the response. Membrane proteins are difficult to extract and are often processed during activation or extraction. In the above example, it could be the case that PlGF activates VEGFR-2, but it is immediately cleaved and internalized. If the antibody that is used only detects the extracellular portion of the protein, whereas it is the intracellular portion that phosphorylated, then activation may be missed.

14.3.4 Tagged proteins

Some proteins do not have good antibodies available for them, so recombinant proteins can be expressed with tags on them to facilitate IP. These include haemagglutin, GST and GFP (useful, as it can also show localization of the protein in real time within cells). The major point of concern with these is that the experiments need to be performed using cells that over-express the protein, usually under control of a powerful viral promoter (see Primer 19). Over-expression of a protein may cause it to associate with proteins with which it has no functional association under normal conditions. The best solution for this is for the protein to be expressed under its own promoter, rather than a viral promoter.

14.4 Complementary and/or adjunct techniques

Primer 12: Antibody production.

Primer 13: Immunocytochemistry.

Primer 15: Immunoblotting.

Primer 24: Proteomics.

Further reading and resources

Amin, E.M., Oltean, S., Hua, J., Gammons, M.V., Hamdollah-Zadeh, M., Welsh, G.I., Cheung, M.K., Ni, L., Kase, S., Rennel, E.S., Symonds, K.E., Nowak, D.G., Royer-Pokora, B., Saleem, M.A., Hagiwara, M., Schumacher, V.A., Harper, S.J., Hinton, D.R., Bates, D.O. & Ladomery, M.R. (2011). Alternative VEGF splicing in WT1 mutants reveals SRPK1 to be a novel angiogenesis target. *Cancer Cell* **20**(6), 768–80.

Masters, S.C. (2004). Co-immunoprecipitation from transfected cells. *Methods In Molecular Biology* **261**, 337–50.

Warner, A.J., Lopez-Dee, J., Knight, E.L., Feramisco, J.R. & Prigent, S.A. (2000). The Shc-related adaptor protein, Sck, forms a complex with the vascular-endothelial-growth-factor receptor KDR in transfected cells. *Biochemical Journal* **15**, 347(Pt 2) 501–9.

15
Immunoblotting (Western)

Samantha F. Moore, Joshua S. Savage and Ingeborg Hers

Physiology & Pharmacology, University of Bristol, UK

15.1 Basic 'how-to-do' and 'why-do' section

Immunoblotting (aka Western blotting) is used for the transfer of proteins in samples such as cell lysates, tissue homogenates and immunoprecipitates (see Primer 14) from a gel to a substrate or membrane, and their subsequent detection using antibodies on the surface of the membrane. The specificity of the antibody-antigen interaction means that a single protein can be identified in a complex protein mixture and information about its size and phosphorylation status can be obtained. Immunoblotting therefore allows us to:

- verify the selectivity of antibodies (see Primer 12);
- assess whether or not cells or tissues express a particular protein;
- determine whether the expression level of the protein changes under altered experimental conditions;
- determine protein modification (e.g. phosphorylation, ubiquitination, lipoylation, glycosylation);
- determine the activation state of a protein (e.g. small G-proteins).

15.1.1 Protein separation

The first step in an immunoblot procedure is to separate the proteins according to their approximate molecular weights using SDS-polyacrylamide gel electrophoresis (PAGE) (Figure 15.1a). The principle of gel electrophoresis is that proteins are separated according to their negative charge using an electric field.

Essential Guide to Reading Biomedical Papers: Recognising and Interpreting Best Practice, First Edition.
Edited by Phil Langton.
© 2013 by John Wiley & Sons, Ltd. Published 2013 by John Wiley & Sons, Ltd.

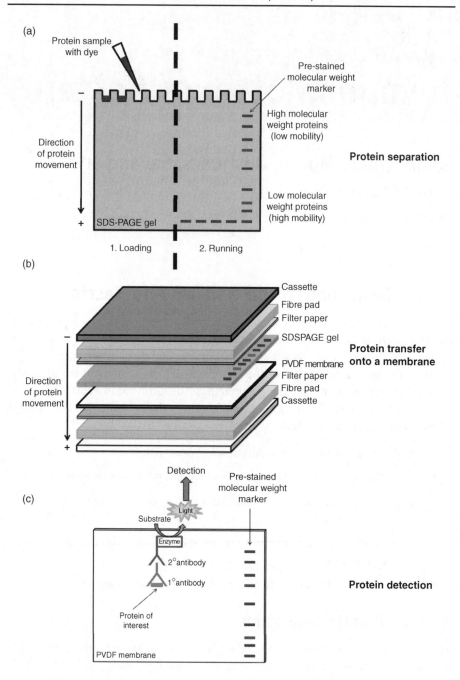

Figure 15.1 *(Continued on next page)*

To allow proteins to be separated according to size, the sample is first treated with SDS, a strong reducing agent that denatures proteins and binds to the resulting polypeptides. This gives the protein (polypeptide) a uniform charge with a constant charge-to-mass ratio and allows separation according to size. A tracker dye with a higher electrophoretic mobility than proteins is also added to the sample, and this allows the researcher to estimate when the electrophoresis is finished (when the dye reaches the bottom of the gel).

The gel matrix used is polyacrylamide. Different percentage SDS-PAGE gels and different gel buffers can be used, and the choice depends on the ultimate research question. The most commonly used SDS-PAGE gels are Tris-glycine based gels, but better resolution and sharper bands are often obtained with Bis-Tris based gels. Tris-glycine gels, however, are still the choice when looking for subtle protein band shifts or subtle differences in molecular weight. Pre-stained molecular weight markers are run in order to estimate the apparent molecular weight of the detected bands/proteins.

15.1.2   Protein transfer onto a membrane

Following electrophoresis, the separated proteins are transferred onto a membrane (see Figure 15.1b), which is generally made of nitrocellulose or polyvinylidene (PVDF). PVDF is more resilient and has a higher binding capacity compared to nitrocellulose, but it requires more stringent blocking and may produce higher background for certain antibodies. The membrane is positively charged and binds SDS-bound proteins non-specifically, so that all proteins in the sample can be exposed to the antibody.

The transfer method that is most commonly used is electro-elution, either by wet or semi-dry transfer, which is relatively quick (1–2 hours) and efficient. The gel is placed in direct contact with the membrane and is sandwiched between filter papers

◄──

(*Continued*) Principles of immunoblotting. (a) Protein separation: the protein sample is mixed with SDS and a dye and loaded on a SDS-PAGE gel. The SDS gives the protein (polypeptide) a uniform charge, which allows separation according to size. When a current is applied, the negatively charged proteins move in the direction of the positive (anode) pole. High molecular weight proteins move slowly and remain near the top of the gel, whereas low molecular weight proteins move faster and end up near the bottom of the gel. Pre-stained molecular weight markers (containing proteins of known molecular weight) are also run on the gel in order to estimate the apparent molecular weight of proteins detected in C. (b) Protein transfer onto a membrane: proteins are transferred from the gel onto a membrane by electro-elution. The negatively charged proteins move out of the gel towards the positive pole and bind to the membrane. During this procedure, the gel and membrane are in close contact and submerged in transfer buffer. (c) Protein detection; the membrane is incubated with primary (1st) antibody against the protein of interest, followed by a enzyme labelled secondary (2nd) antibody that binds to the primary antibody. Enzymatic substrate conversion results in the production of light (chemiluminescence), which can be detected by X-ray film. Alternatively, fluorescently labelled secondary antibodies can be used. Courtesy of Dr. Ingeborg Hers. *A full colour version of this figure appears in the colour plate section.*

and two electrodes submerged in a conducting solution (Figure 15.1b). When an electric field is applied, the proteins move out of the gel and onto the surface of the membrane. The efficiency of transfer depends on many factors and relies on complete contact between the gel and the membrane. Air bubbles, for example, between the gel and the membrane, will insulate that portion of the gel and so block protein transfer.

The transfer of protein can be checked by eye (after transfer, leave the PDVF membrane on the wet filter paper to semi-dry and view the membrane from a 45 degree angle; bands should be visible on PDVF membrane) or by staining the membrane with a reversible reagent such as ponceau S solution. Complete transfer of the pre-stained markers is also a good indication of protein transfer. The membrane is subsequently '*blocked*' with a buffer containing proteins ranging from dry milk, Bovine serum albumin or normal serum, for a few hours at room temperature or overnight at 4 °C. Blocking is used to prevent non-specific binding of antibodies to the membrane during subsequent steps in the assay, and it should improve sensitivity by reducing background staining without obscuring the epitope for the antibody binding.

Some blocking buffers, however, are not compatible with every system. Milk-based protein blocking buffers interfere with avidin-biotin containing systems, because milk contains variable amounts of biotin. Similarly, milk often causes very high background when using anti-phosphotyrosine antibodies.

15.1.3 Detection

Like other immunoassay procedures, such as immunocytochemistry, immunoblotting consists of a series of antibody incubations separated by wash steps (Figure 15.1c). Washing the membrane between steps removes unbound reagents and reduces background staining of the membrane, increasing the signal-to-noise ratio. Washing is performed in a buffered solution such as Tris-buffered saline, usually with a detergent such as Tween 20, which helps remove non-specifically bound material.

The presence of the protein of interest is detected using a primary antibody, either polyclonal or monoclonal (see Primer 12). Generally, the antibody is diluted into buffer containing detergent and diluted blocking solution, and incubated with the blot for 1–2 hours at room temperature. It will usually have been necessary to optimize the antibody concentration. If the antibody concentration is too low, no signal will be detected; if it is too high, non-specific bands may appear.

The primary antibody is then detected using a secondary antibody raised against the species of the animal in which the primary antibody was raised. For example, if the primary antibody is a rabbit polyclonal, then the secondary antibody must be an anti-rabbit antibody from a host other than a rabbit. The choice of secondary antibody depends on the type of label required. Enzymatic labels such as alkaline phosphatase (AP) and horseradish peroxidase (HRP) are extremely sensitive and are used extensively.

Secondary antibody-HRP (or -AP) conjugates can then be detected using colour, fluorescent and chemiluminescent substrates. The chemiluminescent substrates are the most popular; in the presence of HRP and H_2O_2, they emit light that can be detected by X-ray film. Bands corresponding to the detected protein of interest will appear as dark regions on the developed film. Fluorescently labelled secondary antibodies can also be used to detect the protein of interest, using specialized imaging equipment for visualization.

15.1.4   Densitometry

Quantification of the bands detected by immunoblotting is useful when comparing protein expression levels and/or phosphorylation levels of proteins. When chemiluminescent substrates and X-ray films are used, bands can be quantified by scanning the film, followed by quantification of the intensity of the band using freeware programs such as *ImageJ* (http://rsbweb.nih.gov/ij/).

The intensity of phosphorylated bands can be directly compared to the intensity of the total reprobe for the protein of interest (see 15.3.2). The relationship between protein concentration and band intensity, however, shows poor linearity, and this method therefore represents only a semi-quantitative measure. If quantification is essential to address a research question, it will be beneficial to use a fluorescent substrate approach that demonstrates a better and wider range of linearity. An example is the infrared imaging detection method by Odyssey (LI-COR Biosciences), which uses fluorescently labelled secondary antibodies. This approach has the added advantage of simultaneous detection of fluorescence two different wavelengths (using different labelled secondary antibodies), and therefore allows immunoblotting for two different proteins or targets on the same membrane.

15.2   Required controls

15.2.1   Controls for immunoblotting test the specificity of the antibody-antigen interaction

Nowadays, most primary antibodies are obtained from commercial sources, but it should not be assumed that the antibodies are therefore specific and properly tested (see Primer 12). Information on specificity is not usually available to the reader of a published paper and, in general, key controls relating to antibody specificity tend to be absent. If a single band of correct molecular weight (size approximations are done by comparing stained band with that of a pre-stained protein molecular weight marker) is detected, one might believe the antibody specifically binds the protein of interest. In published papers, however, a select region of the blot at the correct molecular weight is usually shown, with no indication where the molecular weight markers run. This probably reflects space constraints for publication, but it makes it difficult for the reader to assess primary antibody specificity, since other regions of the blot and proper controls are not shown.

Characterization of a primary antibody for use in immunoblotting should include some of the following controls:

1. Blocking antibody-antigen interaction with the peptide the antibody was raised against – does the band go away?

2. Removing or denaturing the primary antibody – does the band go away?

3. Is there a signal from cells known *not* to express the protein of interest (negative control)?

4. Is the band/signal boosted by over-expression of the protein the antibody is raised against?

5. Is a band seen using a cell or tissue that is known to express the protein (positive control)?

6. Does a different antibody raised against a different epitope on the same protein recognize the same band?

7. Protein band is removed and the protein identified by mass spec (absolute verification – see Primer 24).

15.3 Common problems and pitfalls in execution or interpretation

1. The presence of a protein band at the expected height is usually *not* sufficient evidence to conclude that a protein is expressed in a tissue/cell type. Unless the protein and antibody are very well characterized in the tissue/cell type studied (and described in previous publications), it will be necessary initially to confirm these finding by immunoprecipitation using stringent cell extraction conditions (to ensure full protein extraction and reducing the probability of protein-protein complexes – see Primer 14). The protein of interest is immunoprecipitated with a primary antibody (known to work for immuno-precipitation) and run on a gel, followed by immunoblotting. It is important to use a different antibody to the protein of interest for immunoblotting than used for immunoprecipitation, and *ideally* the antibody is also raised in a different host and against a different epitope. This approach is invaluable in obtaining more confidence that the band being detected is, indeed, the protein of interest. Figure 15.2a shows an example of a false positive.

2. When comparing differences in protein expression between different exper-imental conditions, a 'loading control' should be included. The best way to assess protein loading is to strip and re-probe the blot for the [1]*house-keeping* proteins such as tubulin or actin. Alternatively, the same samples can be loaded onto a separate gel and blotted and stained for either of these proteins.

[1] '*House-keeping*' proteins are so named because it is thought (sometimes with good reason) that all cells will express these proteins, and at levels that do not alter much from tissue to tissue.

3. When phospho-specific antibodies are used, it will be important to re-probe the membrane with an antibody against the phosphorylated protein itself. This will not only act as a loading control, but will also confirm that the band recognized by the phospho-specific is, indeed, the protein of interest (*as bands will exactly overlap*). An example is shown in Figure 15.2b.

4. Complete extraction of proteins is essential for interpretation of protein expression data. For example, if a protein translocates to the membrane upon cellular stimulation, it may become more difficult to extract (see Primer 14). Conversely, a membrane-bound protein that becomes cytosolic upon cellular stimulation becomes easier to extract. These results may be wrongly interpreted as changes in protein expression (see Figure 15.2c).

5. Antibodies may recognize the same protein in different tissues with different affinities (possibly due to differences in post-translational modification). This possibility should be considered when protein expression levels between tissues are compared. Ideally, more than one antibody directed against different epitopes on the same protein should be used (Figure 15.2d).

6. Similarly, the ability of an antibody to recognize a protein may be reduced by post-translational modification such as phosphorylation, in particular when the epitope against which the antibody is raised is in the proximity of phosphorylation sites. This is often first noticed when re-probing phosphorylated proteins with a total antibody (Figure 15.2e).

7. The height of the protein band may not be the same for different tissues, cell types and species. Every tissue/cell type has its own repertoire of proteins and may, therefore, show subtle differences in running pattern on SDS PAGE. Highly abundant proteins, for example, may significantly affect the running of less abundant proteins in their vicinity. Differences in post-translational modification between tissues may also affect the apparent molecular weight of the protein band (Figure 15.2f).

8. Over-exposure of the X-ray film to chemiluminescence can lead to 'over-exposed' black bands that do not allow comparison of protein or phosphorylation levels between different samples (Figure 15.2g). These should, therefore, be avoided in publications. If less protein is loaded onto the gel, less primary or secondary antibody is used and/or the X-ray film is exposed for a shorter time, semi-quantitative comparison can be made of the protein of interest from different samples.

9. Similarly, bands may appear on X-ray film as white bands, with or without a black border (see Figure 15.2h). This happens when the chemiluminescent substrate is used up, and therefore indicates a very high signal. This can usually be avoided by reducing the primary and/or secondary antibody concentration. Blots showing whiteouts should never be published.

10. If multiple bands appear on the blot, the specificity of primary antibody may be poor, making it doubtful that the band at the correct height is the protein of interest (Figure 15.2i). The identity of the band will, therefore, need to be confirmed by using the appropriate controls and using the approach described under point 1 (above). However, signal strength may become a problem if the non-specific band(s) are of stronger intensity. If alternative antibodies are not available, the signal intensity of the specific band can be boosted by cutting off the part of the membrane which shows non-specific bands. This avoids the antibody being soaked up by non-specific proteins (Figure 15.2j).

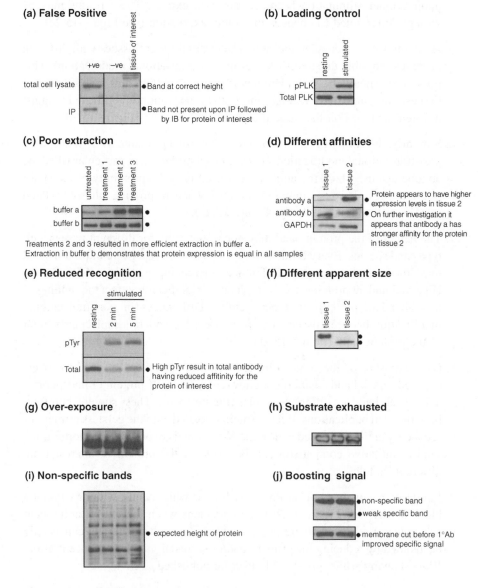

Figure 15.2   *(Continued on next page)*

15.4   Complementary techniques

- Primer 12: Antibody production.

- Primer 13: Immunocytochemistry.

- Primer 14: Immunopreciption.

Further reading and resources

Qiagen Bench Guide online: www.qiagen.com/literature/BenchGuide/pdf/1017778_Bench
   Guide.pdf

LI-COR Quantitative Immunoblotting: www.licor.com/bio/applications/applications.jsp

◄ ───

(*Continued*) Problems and pitfalls in execution or interpretation of Western blots. (a) False
positive: the protein of interest was blotted for in total cell lysate and a band at the correct
molecular weight was observed in the tissue of interest. However, on immunoprecipitation (IP) of
the protein and blotting for the protein with a different antibody, the band was not detected. The
black line in the blot indicates two lanes that have been removed from the image to improve clarity.
(b) Loading control: the phosphorylated form of pleckstrin (PLK) was detected in the stimulated
sample, and the membrane was stripped and re-probed for total pleckstrin to ensure that similar
levels of protein were loaded for each sample. (c) Poor extraction: cells were untreated or treated
with three different treatments and lysed in buffer *a*, and it appeared that treatments 3 and 4
increased the amount of protein of interest. When buffer *b* was used to lyse the cells, no change in
the amount of protein of interested was detected between treatments. Treatments 3 and 4 therefore
resulted in better extraction of the protein in buffer *a*. (d) Different affinities: using antibody *a*, it
appeared that tissue 2 had a greater expression of the protein of interest. Using antibody *b*,
however, demonstrated that tissue 1 and 2 had similar levels of the protein. (e) Reduced
recognition: stimulating cells resulted in tyrosine phosphorylation (pTyr) of the protein; when
probing for the total level of the protein, it was observed that the total antibody had reduced
affinity for the phosphorylated form. (f) Different size: tissue 1 and tissue 2 both express the
protein of interest. However, the protein runs at different molecular weights in the two tissues. (g)
Over-exposure: the bands from this immunoblot are over-exposed, making a comparison of protein
levels between different samples impossible. (h) Substrate exhausted: the bands on this immunoblot are white, due to the chemiluminescent substrate being used up. (i) Non-specific bands:
detection of a ladder of bands. (j) Boosting signal: the top immunoblot shows that the antibody
binds to a non-specific band with high affinity, resulting in the antibody being swamped and the
specific band being picked up only weakly. Cutting the membrane to remove the non-specific band
before incubation with the antibody resulted in an improved specific signal. Courtesy of Dr.
Ingeborg Hers.

16
Applications of Green Fluorescent Protein (GFP)

Mark Jepson

Physiology & Pharmacology, University of Bristol, UK

16.1   Basic 'how-to-do' and 'why-do' section

16.1.1   Fluorescent proteins and their diverse applications

Green **F**luorescent **P**rotein (GFP), from the jellyfish *Aequorea victoria*, has revolutionized cell biology. GFP, linked to proteins or peptides and expressed in cells, is used extensively to reveal the location of proteins, lipids and sub-cellular compartments, and to analyze protein-protein interactions. GFP is also used widely to measure gene expression and to locate expressing cells and organisms in applications ranging from viruses to model organisms, intact mammals and plants. Engineering of the native GFP protein has generated many mutant forms that are brighter and more stable (e.g. EGFP), or have different spectral properties (e.g. CFP, YFP, BFP), allowing multiple labelling experiments.

Where GFP is used to monitor changes in promoter activity, destabilized GFP variants may provide a more accurate transcriptional reporter system. Photo-activatable GFP variant (PA-GFP) allows protein within part of the cell to be photolabelled by selective illumination in order to monitor protein trafficking. Other GFP variants (pHluorins) are pH-sensitive, becoming fluorescent at higher pH. Constructs comprising two GFP variants (usually CFP and YFP) with overlapping spectra have also been used as targeted reporters of Ca^{2+} concentration and enzyme activity (see below).

Fluorescent proteins from other marine organisms have now been added to the extended fluorescent protein family. For example DsRed, from reef coral, is a red fluorescent protein providing fluorescence at wavelengths far removed from GFP

Essential Guide to Reading Biomedical Papers: Recognising and Interpreting Best Practice, First Edition. Edited by Phil Langton.
© 2013 by John Wiley & Sons, Ltd. Published 2013 by John Wiley & Sons, Ltd.

variants. During maturation, DsRed goes through a transient green fluorescent form, and a mutant form of DsRed aims to exploit this property to provide a measure of transient promoter activity. Another coral protein, Kaede, undergoes UV-dependent photo-conversion from a green to a red fluorescent form and is used as an optical marker of cells. Yet another red fluorescent protein, KillerRed, is used to kill expressing cells selectively due to its extreme phtototoxic properties.

While this provides a tiny glimpse of the dazzling array of fluorescent proteins now available and their potential applications, it is neither possible here to provide a comprehensive overview of this rapidly developing field, nor to discuss the molecular biology techniques required to exploit the fluorescent protein toolkit. It is, nevertheless, worth discussing some of the specialized techniques that are being used in conjunction with GFP technology.

16.1.2   Advanced fluorescence techniques exploiting GFP and its variants

A number of techniques are used to monitor mobility of GFP-labelled proteins and their interactions in cells. These may not always be explained clearly in research papers and are often referred to using potentially confusing acronyms.

While the mobility of discrete compartmentalized structures containing GFP-labelled proteins can be monitored with simple live cell (time-lapse) imaging, this technique is often insufficient to monitor the movement of molecules within and between cellular compartments. Subsidiary techniques use photobleaching (irreversible loss of signal due to light-induced fluorophore damage) or photo-activation (light-induced switching of fluorescent properties) within selected areas of cells to monitor protein mobility in more detail. FRAP (**F**luorescence **R**ecovery **A**fter **P**hotobleaching) monitors movement of fluorescent molecules into an area that has been bleached by high intensity illumination – usually by a laser. FLIP (**F**luorescence **L**oss **I**n **P**hotobleaching) is related to FRAP, but monitors loss of fluorescence in some regions of the cell, while another area is repeatedly bleached. If distant fluorescence is reduced, FLIP can determine from where fluorescence molecules migrate into the area being photo-bleached.

These techniques (FRAP and FLIP) should not to be confused with FLAP (**F**luorescent **L**ocalization **A**fter **P**hotobleaching), which selectively labels a sub-population of proteins, allowing its subsequent movement to be traced. In FLAP, protein is labelled with two alternative fluorophores (e.g. CFP and YFP), then one of these is selectively photobleached within a discrete region to label the population of protein molecules in that region with a specific fluorescence signature. These techniques for monitoring protein mobility have now been joined by photoactivatable and photoswitchable GFP that change their fluorescence properties following illumination. As with FLAP, but more simply, these facilitate selective labelling of cells or proteins within a region of the cell and subsequent tracing by monitoring the activated GFP.

Protein-protein interactions can be monitored using 'FRET' – **F**luorescence (or **F**örster) **R**esonance **E**nergy **T**ransfer. FRET is the transfer of energy between any pair of fluorophores ('donor' and 'acceptor'), and it only occurs when the fluorophores are very close together (normally within 10 nanometres) and have overlapping emission and excitation spectra. Some energy that would normally be released as light by the donor is directly transferred to the acceptor, which is thereby excited and emits light at its normal emission wavelength. FRET can detect interactions between proteins labelled with spectral variants of GFP with overlapping fluorescent spectra (e.g. CFP and YFP).

FRET has also been used to monitor cell signalling using FRET pairs (e.g. CFP-YFP) linked by amino acid sequences which can undergo structural movement to bring together or separate the fluorophores, changing their FRET efficiency. Such paired constructs have been used to monitor site-specific phosphorylation, Ca^{2+} fluxes and protease activity, sometimes targeted to cellular compartments.

16.2 Required controls

Some basic controls (negative and positive) should be used in experiments using fluorescent proteins to monitor their expression in cells. Explanation of this can be found in Primer 8. The analysis of photobleaching experiments with fluorescent proteins also requires careful interpretation, e.g. confirmation that FRAP 'recovery' does not occur in fixed samples, tests to ensure that bleaching is not damaging cellular machinery, etc. These are rather specialized areas and there is not sufficient space to discuss details here.

16.2.1 Verifying correct behaviour of GFP-labelled proteins

The expression of GFP-labelled proteins in cells to study the localization and dynamics of proteins relies on GFP-labelled protein chimeras behaving like the native proteins. In most examples, GFP-labelled proteins behave normally, but it may be important to perform controls to ensure that this is so.

The location of endogenous protein is often confirmed by immunofluorescence to demonstrate that an exogenously expressed GFP-tagged version is correctly localized. Researchers might confirm that the GFP-labelled protein functions normally by knocking out endogenous protein expression and then showing that its function is substituted by expression of a GFP-labelled version, but this control is not always feasible or appropriate. In the absence of such controls, it follows that the researchers are making an assumption that the GFP-labelled protein has normal, or at least adequate, function. Accumulation of GFP-labelled protein in a particular compartment may be verified by negative controls. Here, cells are transfected with a similar plasmid expressing native GFP, or an alternative or modified GFP-construct that should not show similar enrichment in the cellular compartment (see Primer 19).

16.2.2   Controls for fluorescence resonance energy transfer (FRET) experiments

FRET imaging studies require very careful verification with appropriate controls. Using FRET to analyze protein-protein interaction in live cells is especially difficult. Part of the difficulty arises because this requires co-transfection of two fluorescent-labelled proteins, and these are likely to be expressed in variable amounts or to be compartmentalized in an uneven manner within cells. Difficulties in interpretation of fluorescence signal arise due to spectral overlap, which means that the FRET signal is always contaminated by bleedthrough of donor emission into the acceptor channel and by the excitation of acceptor molecules by the donor's excitation wavelength.

The situation is simplified for FRET studies using a single linked construct (to analyze Ca^{2+}, kinase and protease activity, etc.) because the FRET donor and acceptor are always co-localized and, hence, the cross-talk between their signal will be a constant factor, enabling reliable FRET measurements. For standard FRET imaging, it is necessary to measure the degree of cross-talk between both excitation and emissions to accurately calculate FRET efficiency. This is usually done by carefully controlled imaging of the acceptor and donor in isolation. Reliable FRET imaging depends on accurate detection of relatively small increments of fluorescent signal compared to baseline signal levels and, for this reason, it can be technically demanding.

Subsidiary experiments are also often performed to confirm that a genuine FRET signal is being measured. Bleaching the FRET acceptor destroys its ability to accept energy from the donor, leading to loss of FRET signal as well as an increase in emission at the donor's emission wavelength. Similarly, a FRET donor is partially protected from bleaching if it transfers energy to a FRET acceptor, so FRET efficiency is related to the rate of bleaching of the donor.

The time a fluorophore remains in its excited state (nanosecond scale) is reduced when FRET occurs, and this can be monitored with sophisticated microscope systems (**F**luorescence **L**ifetime **I**maging **M**icroscopy – FLIM), using rapidly pulsed excitation and accurate detection. Measuring the decreased fluorescence lifetime of FRET donors provides very sensitive analysis, but FLIM systems are complex and yet to become widely available.

As with other applications, it may also be helpful to have positive and negative controls. FRET can be challenging to record, which can lead to errors of interpretation (e.g. absence of evidence being interpreted as evidence of absence). To address this, if no FRET signal is detected in studies using FRET to detect or measure protein-protein interaction, it can be reassuring if a positive signal from an alternative protein pair, or physically linked donor/acceptor, is presented. Similarly where FRET constructs are used as reporters of ion fluxes or changes in enzyme activity, it is helpful to compare a treatment that reproducibly induces a change in FRET signal. Where protein-protein interaction is measured by FRET, it can also be helpful to compare signal from a similar pair of proteins where the fluorophores do not interact. In some studies, this type of negative control has been performed by comparing protein constructs, in which the fluorescent proteins are displaced to prevent their interaction.

16.3   Common problems and pitfalls in execution or interpretation

Many potential problems and pitfalls in studies with fluorescent proteins are general microscopy issues covered in the Primer 8 (Fluorescence microscopy) or have been discussed in the sections above.

A major issue in working with GFP (and other fluorescent protein) chimeras is verification that the presence of the fluorescent protein does not affect the behaviour of the protein it is linked to. As we have seen, this appears to be normally the case with GFP-labelled proteins, but is nevertheless something that needs to be carefully considered in all cases. Early studies with fluorescent protein alternatives to GFP – especially red fluorescent versions – were prone to problems with aggregation and, in some cases, with changing spectral properties during maturation. These problems have been overcome with monomeric versions and new spectral variants such as the 'mFruit' proteins (mCherry, mPlum etc).

There are many cases where over-expression of fluorescent protein chimeras interferes with normal function of the protein, e.g. by binding to interacting proteins. Usually, transfection conditions are selected to achieve a level of expression that reduces potential problems of aberrant localization or behaviour associated with over-expression. In practice, the levels of expression achieved are often variable, and it is common practice to select cells to examine that have 'good' levels of expression of the GFP-construct, i.e. are sufficiently fluorescent to enable imaging while avoiding over-expression artefacts.

Transfection agents also exhibit varying levels of toxicity that differ between cell-lines and culture conditions; again, conditions need to be selected to minimize toxic effects on cells. Virus infection is increasingly being used to transfer fluorescent proteins more stably and reproducibly into cells, particularly those that are difficult to transfect using plasmids. These issues will be discussed in more detail elsewhere (see viral vector transgenesis, Primer 19).

16.4   Complementary techniques

Primer 8: Fluorescence microscopy.
Primer 13: Immunocytochemistry.
Primer 19: Viral vector transfection.
Primer 22: Methods of interference.

Further reading and resources

www.bdbiosciences.com/clontech/gfp/index.shtml (Commercial website detailing the extensive range of fluorescent proteins sold by Clontech).

Day, R.N. & Davidson, M.W. (2009). The fluorescent protein palette: tools for cellular imaging. *Chemical Society Reviews* **38**, 2887–2921 (Extensive review covering the range of fluorescent proteins and imaging methodology).

Giepmans, B.N., Adams, S.R., Ellisman, M.H. & Tsien, R.Y. (2006). The fluorescent protein toolbox for assessing protein location and function. *Science* **312**, 217–224 (Overview of applications of GFP and alternative fluorescent labelling techniques).

Perestenko, P.V. & Henley, J.M. (2003). Characterization of the intracellular transport of GluR1 and GluR2 alpha-amino-3-hydroxy-5-methyl-4-isoxazole propionic acid receptor subunits in hippocampal neurons. *Journal of Biological Chemistry* **278**, 43525–43532 (Research article includes GFP trafficking, pH-sensitive GFP, FRAP, FLIP data).

Timpson, P., McGhee, E.J. & Anderson, K.I. (2011). Imaging molecular dynamics in vivo – from cell biology to animal models. *Journal of Cell Science* **124**, 2877–90. (Overview of fluorescent protein imaging applications, emphasizing considerations required for in vivo studies)

17
Fluorescent Measurement of Ion Activity in Cells

Helen Kennedy

Physiology & Pharmacology, University of Bristol, UK

17.1   Basic 'how-to-do' and 'why-do' section

This technique is generally used to measure changes in ion concentrations in living cells. Many ion species, such as calcium, chloride, magnesium and sodium, play important physiological roles; to learn more about these processes, it is important to be able to measure changes in their concentration in living cells.

17.1.1   General characteristics of fluorescence

A fluorescent molecule can absorb radiation (light) at one wavelength, and the resulting excitation of the molecule leads to emission of the radiation (light) at a longer wavelength. To be useful, fluorescent molecules (sometimes called probes or indicators) should have the following properties:

1. A change in fluorescent properties in response to binding the ion of interest, which is of sufficient intensity to produce a good signal-to-noise ratio for measurement. Such changes can be monitored by the experimenter and related to changes in ion concentration.

2. High specificity for the ion of interest, and therefore little or no interaction with other ions.

3. The ability to detect the ion at physiological concentrations. This is related to the affinity (often called K_d) of the molecule. When the K_d is very low, it is a *high-affinity molecule* and can measure ions that exist at very low concentrations.

Essential Guide to Reading Biomedical Papers: Recognising and Interpreting Best Practice, First Edition. Edited by Phil Langton.
© 2013 by John Wiley & Sons, Ltd. Published 2013 by John Wiley & Sons, Ltd.

However, these dyes are easily saturated. If K_d is high (micromole order of magnitude or bigger), we talk about *low-affinity* molecules that are used to measure ions present at higher concentrations but are poor at detecting small changes in the ion of interest. Ions that undergo large dynamic changes in concentration, such as calcium, often require the use of several related dyes with different K_d to measure changes accurately over the whole dynamic range.

17.2 Methodology for measuring ion concentrations

a) **Fluorescent dyes:** one of the most common and widely used methods for measuring ion concentrations in living cells is the use of fluorescent dyes. These dyes are widely available for a variety of intracellular ions and often available in versions with different K_d. These variants in K_d can be particularly useful in studying localized calcium signals, such as those shown in Figure 17.1, imaged using a low affinity dye.

b) **Molecular approaches:** more recently, a more genetic approach to ion measurements has become more widely available. These fluorescence techniques are based on the GFP-based sensor family, and they rely on the expression of genetically encoded sensors. Most often, these sensors are based on two fluorescent proteins joined by the ion sensor. Upon ion binding, a conformational change occurs that brings the two fluorescent protein domains closer, allowing fluorescence resonance energy transfer (FRET) to occur (see Primer 16).

17.2.1 Measuring ions with fluorescent dyes

Non-ratiometric dyes increase their fluorescence intensity (intensity shift) upon binding the ion. A commonly used single-excitation, single-emission dye for

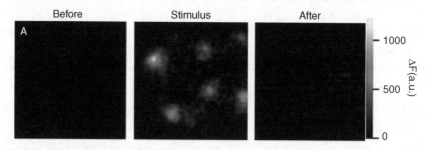

Figure 17.1 Ca^{2+} microdomains mediated by Ca^{2+} influx at ribbon synapses in IHCs. Confocal images of Fluo-5N-filled IHCs were acquired before, during, and after stimulation by 200 ms depolarizations to −7 mV. Images averaged over runs and time frame are shown. (Scale bars: 2 μm.) Adapted from Thomas Frank, Darina Khimich Andreas Neef, and Tobias Moser. (2009) Mechanisms contributing to synaptic Ca^{2+} signals and their heterogeneity in hair cells PNAS 106 (11): 4483–4488. © National Academy of Science. *A full colour version of this figure appears in the colour plate section.*

measuring intracellular calcium is fluo-3, and there are several closely related dyes based on this molecule with varying properties. This molecule, like many other non-ratiometric dyes, can be excited efficiently with light at 488 nm and exhibits minimal fluorescence when calcium is low. Upon binding calcium, it undergoes a dramatic increase in fluorescence emission at 525 nm, giving excellent signal-to-noise ratio.

Because many non-ratiometric dyes are available in forms with absorption maxima closely associated with common laser spectral lines, they are often the dyes of choice for confocal microscopy. The combination of fluorescent dyes and confocal microscopy allows imaging of both global and localized calcium increases, such as those on dendritic spines, shown in Figure 17.2.

The disadvantage of non-ratiometric measurements is that changes in fluorescence intensity can be caused by changes in dye concentration as well in the ion of interest. If the dye is unevenly distributed within the cell, or there is any loss or photobleaching, it may 'appear' that the concentration of the ion is changing. One approach to this problem is to use combinations of dyes with different spectral properties to produce *pseudo-ratiometric* measurements, although this approach in itself is not without its disadvantages, such as increased buffering and potentially increased toxicity.

Ratiometric dyes undergo either a shift in emission or excitation wavelength. A ratiometric measurement can be made by using a ratio between these two fluorescence intensities (either the emission or excitation wavelengths). The major advantage of these dyes is that because the fluorescence signals are ratiometric, they are not related to dye concentration, making the signals much easier to interpret.

Unfortunately, for those experimenters who wish to use confocal microscopy to obtain increased spatial information, dyes excited by UV light present some problems. Although confocal microscopes can be fitted with UV lasers, they are costly, and UV light gives limited specimen penetration depths and produces optical aberrations. These drawbacks mean that ratiometric UV-based dyes are not commonly used in confocal microscopy.

17.2.2 Different dye forms

Fluorescent dyes are available in one or all of the following forms:

1. Cell impermeant indicators, usually in the form of a potassium or sodium salt. An advantage of this type of dye is that it is much less likely to accumulate in intracellular compartments. The disadvantage is that it must be loaded into the cell directly, which risks cell damage. One common method is to load the indicator through a patch pipette during whole-cell recording. Other methods include microinjection or electroporation.

2. Cell permeable indicators (AM esters). These dyes are esterified by having acetoxymethyl (AM) groups attached, making the dye lipid-soluble so that it can readily cross the cell membrane. Chemicals such as Pluronic F-127 can be

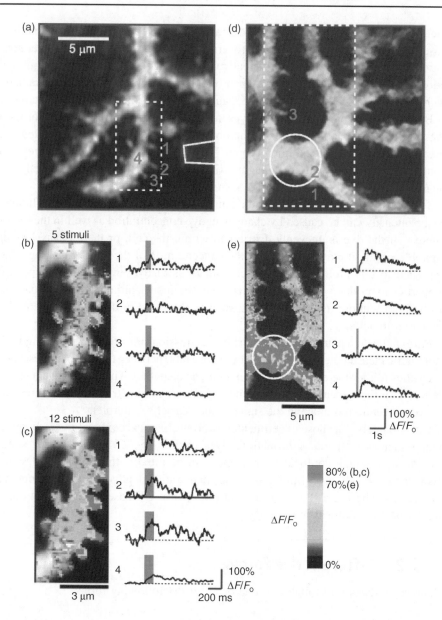

Figure 17.2   InsP3-mediated Ca^{2+} signals produced in Purkinje cell dendritic spines and shafts by parallel-fibre (PF) activity (left) and uncaging InsP3 (right). The area studied in (b, c) (outlined by dashed box) and the position of the PF stimulation pipette are shown. Numbers indicate locations from which Ca^{2+} signals were measured in (b, c). (b): Ca^{2+} signals produced by 5 stimuli (50 Hz) in CNQX. Traces represent Ca^{2+} signals in the numbered regions of (a). (c): Ca^{2+} signals produced by 12 stimuli (80 Hz). (d): Circle indicates position of the ultraviolet light spot (5 μm diameter) and numbers indicate locations from which Ca^{2+} signals were measured in (e). (e), Ca^{2+} signals produced by uncaging InsP$_3$. Reproduced, with permission, from Elizabeth A. Finch & George J. Augustine (1998) Local calcium signalling by inositol-1,4,5-trisphosphate in Purkinje cell dendrites Nature 396, 753–756 © Nature. *A full colour version of this figure appears in the colour plate section.*

used to encourage thorough dispersion of the dye in the loading solution. Once inside the cell, esterases cleave the AM groups, leaving the charged form of the dye trapped inside the cell. This can be problematical, as some cells are less efficient at cleaving the dyes, leaving poorly de-esterified dyes in the cytoplasm. These molecules are usually unresponsive to the ion of interest, yet still fluorescent.

In addition, care should be taken with loading times, to balance the need for enough dye to achieve a good signal versus putting too much dye into the cell, and so efficiently buffering the ion of interest and altering the kinetics of any changes under study. A common problem with AM-dyes is that they have a tendency to accumulate in intracellular compartments, although this can be minimized by adjusting loading conditions. For some, this represents a disadvantage as it can reduce the quality of the signal-to-noise ratio or render the signal hard to interpret. This can happen, for example, if cytosolic levels rise while those in intracellular compartments fall, or vice versa.

However, some experimenters use the ability of these dyes to accumulate in intracellular compartments as a tool for measuring ion concentrations inside intracellular organelles. An example of this is measuring intracellular calcium stores by using a dye with very low affinity for calcium (e.g. Mag-fluo-4, originally designed to measure magnesium) and loading it under conditions that maximize accumulation into intracellular compartments. Cytosolic dye can then effectively be washed out of the cell using the whole-cell configuration of the patch clamp technique. Any fluorescence changes then reflect changes in ion concentrations in the intracellular compartments such as ER. Such measurements are shown in Figure 17.3.

3. Dextran-bound dyes are based on the membrane impermeant dyes with the addition of a hydrophilic polysaccharides dextran. This produces a dye with high molecular weight that has good water solubility, low toxicity, is relatively inert and is resistant to cleavage by intracellular enzymes. Due to the addition of the high molecular weight dextran, these dyes are not commonly taken up into intracellular organelles or efficiently extruded from the cell, giving them excellent longevity.

17.2.3   How are the measurements made?

Four essential elements of fluorescence detection systems can be identified from the preceding discussion:

1. An excitation source.

2. A fluorophore.

3. Wavelength filters to isolate emission photons from excitation photons.

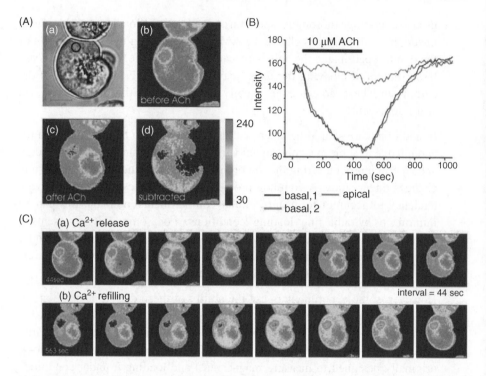

Figure 17.3   Time course of ACh-evoked Ca^{2+} release and the subsequent refilling of the ER. In a Mag-fluo 4-loaded cell, 10 μM ACh was applied. (A): (a) Transmitted light picture; (b and c) fluorescence intensity images before and after ACh; (d) subtracted image (b–c). (B): Time course of ACh-induced fluorescence intensity changes in the two basal regions as well as the apical regions marked in [A(a)]. (C): Time course of Ca^{2+} release following ACh stimulation and subsequent refilling after cessation of stimulation. Reproduced, with permission, from Fig 4 in Myoung Kyu Park, Ole H. Petersen and Alexei V. Tepikin (2000) The endoplasmic reticulum as one continuous Ca^{2+} pool: visualization of rapid Ca^{2+} movements and equilibration. The EMBO Journal 19, 5729 – 5739 © Nature. *A full colour version of this figure appears in the colour plate section.*

4. A detector that registers emission photons and produces a recordable output, usually as an electrical signal or a photographic image.

Regardless of the application, compatibility of these four elements is essential for optimizing fluorescence detection (see Primer 8).

The complexity of systems can vary widely, from very simple excitation/emission systems capable of giving information at the whole-cell level, to complex confocal microscopes that allow high resolution at both the spatial and temporal level.

17.3   Required controls and problems or errors in literature

There are a number of potential problems encountered when using fluorescent ion detection.

17.3.1   Changes in the affinity of the dye

This is a common problem in the literature and can occur with any dye. When using an indicator to measure absolute ion concentrations, it is important to calibrate the fluorescence signals. An important part of this is to know the affinity (K_d) of the dye for the ion. Though known for each dye in a saline solution, the K_d of the dye may change substantially in the intracellular environment. Thus, the change in K_d for the dye inside the cell must be determined.

Many people will use the published K_d for the dye rather than measuring it in their system, or determining it *in vitro* in viscous solutions. This can lead to large errors in the estimation of absolute concentrations of the ions being measured, and can be a problem if absolute ion activity is important. In reality, many experimenters choose not to make absolute measurements of ion concentration, but instead use more qualitative and comparative information.

17.3.2   Changing the intracellular signal being measured

Any indicator that binds ions is acting as a chemical buffer. While this has little effect on the resting or raised ionic concentrations, it can slow the kinetics of such changes. This problem is compounded if it is not possible to know the concentration of the dye inside the cell (e.g. with AM-ester dyes, or when microinjecting the cell with impermeant dyes). It is not possible to overcome the buffering effect fully, as some dye must be introduced into the cell in order to make the measurement; however, most modern dyes have high fluorescence yields (low concentrations give clear signals) and techniques have been developed to allow the concentration of the dye to be estimated.

17.3.3   Background fluorescence

This can take the form of autofluorescence, or from dye that has bound to intracellular constituents such as proteins and is no longer ion-sensitive. To overcome autofluorescence, it is useful to use dyes that operate at wavelengths different from that of the autofluorescence. Appropriate techniques for calibrating the signals can also minimize contamination by these signals.

17.3.4   Compartmentalized dye

This is particularly a problem with the AM dyes as they readily cross cell membranes and therefore can potentially cross not just into the cytosol, but into the intracellular organelles such as endoplasmic reticulum and mitochondria. Lowering the temperature at which cells are loaded sometimes reduces this problem. However, it is often necessary to subtract the fluorescent signal from compartmentalized dye, and this can be achieved by using detergents to release cytosolic dye, leaving the signal from the intracellular structure behind. This can be measured and subtracted from the signal.

17.3.5   Loss of dye

Occasionally, cells that are very efficient at extruding the dye molecules are encountered – a problem most often observed with the membrane permeable dyes. This problem can present as if cells are failing to load with the dye. Several inhibitors are available that reduce or eliminate dye extrusion.

17.3.6   Photobleaching

Consider non-ratiometric dyes. During an experiment, the dye is excited with light and, as the intensity of the fluorescence is related to the intensity of the excitation light, high levels of illumination are used. This can damage or *photobleach* the dye, giving the appearance of dye loss. Additionally, the degradation of the dye may release reactive and damaging molecules. Minimizing both the illumination intensity and exposure time reduces photobleaching. Photobleaching of a non- ratiometric dye results in a decline in signal and could be interpreted as a decrease in the concentration of the ion being measured. The use of ratiometric dyes overcomes this problem.

17.4   Molecular techniques

Molecular approaches to ionic measurements are becoming increasingly common, and they have a number of advantages and disadvantages. The most widely available probe of this type was originally created by Miyawaki *et al.* (1997), based on the expression of two fluorescent proteins – most commonly CFP and YFP, linked by an ion-sensitive molecule. The basis of ion measurement relies on the binding of the ion of interest, producing a conformational change in the linker molecule that allows FRET to occur. For details on FRET measurements, see Primers 8 and 16. Although many different types of constructs have been created by different groups for measuring a variety of intracellular ions, only a small number are available commercially.

17.4.1   Advantages

These probes produce ratiometric measurements of the ion of interest, and they can measure ionic concentrations over a wide physiological range.

The very nature of these probes renders them open to genetic modification for specific purposes. For example, addition of specific promoters can enable cell specific expression of these probes. The ability to infect only one cell type in complex tissues such as the brain represents an enormous advantage over traditional dye-based techniques.

Another advantage of this molecular approach is the potential to create probes that can be specifically targeted to cellular locations (e.g. mitochondria or the cell membrane), allowing high spatial resolution to be obtained. Although some laboratories have created such probes, they are not yet commercially available.

17.4.2 Disadvantages and problems encountered

Being genetically encoded, these probes need to be transfected into cells – something that can present problems in some systems. Although efficient viral transfection is used, this in itself is not a trivial procedure and may need considerable effort in order to obtain satisfactory transfection rates and expression levels sufficient to give an acceptable signal-to-noise ratio (Primer 19). Sufficient expression of the probe requires that tissue can be maintained in an appropriate environment (or *in vivo*) long enough for expression of the proteins to occur, something that is not always possible and is seldom trivial.

As with fluorescent dyes, the ion-binding properties and how these interact or compete with the endogenous molecules must be taken into consideration. For example, when using calmodulin-based probes, there may be interaction between the over-expressed calmodulin present in the cameleon that can interact with native calmodulin or calmodulin-dependent mechanisms.

17.5 Complementary and/or adjunct techniques

Primer 8: Fluorescence microscopy.
Primer 16: GFP.
FRET imaging (in GFP primer).
Confocal imaging.

Further reading and resources

Finch, E.A. & Augustine, G.J. (1998). Local calcium signalling by inositol-1,4,5-trisphosphate in Purkinje cell dendrites. *Nature* **396**, 753–756.

Frank, T., Khimich, D., Neef, A. & Moser, T. (2009). Mechanisms contributing to synaptic Ca^{2+} signals and their heterogeneity in hair cells. *Proceedings of The National Academy of Sciences of The United States of America* **106**(11), 4483–4488.

Goldberg, J.H., Tamas, G., Aronov, D. & Yuste, R. (2003). Calcium microdomains in aspiny dendrites. *Neuron* **40**(4), 807–821.

Miyawaki, A., Llopis, J., Heim, R., McCaffery, J.M., Adams, J.A., Ikura, M. & Tsien, R.Y. (1997). Fluorescent indicators for Ca^{2+} based on green fluorescent proteins and calmodulin. *Nature* **388**(6645), 882–887.

Park, M.K., Petersen, O.H. & Tepikin, A.V. (2000). The endoplasmic reticulum as one continuous Ca^{2+} pool: visualization of rapid Ca^{2+} movements and equilibration. *The EMBO Journal* **19**, 5729–5739.

The Molecular Probes website: www.probes.com/servlets/directory?id1=3 (has detailed information about the properties of commonly used fluorescent indictors).

18
Detection of Exocytosis – Real Time

Anja Teschemacher

Physiology & Pharmacology, University of Bristol, UK

18.1 Basic 'how-to-do' and 'why-do' section

Chemical messengers, such as neurotransmitters and hormones, are often contained in vesicles within excitable cells, and are released into the extracellular space by exocytosis. Detecting exocytotic events in real time allows us to characterize the molecular mechanisms underlying dynamic changes in transmitter release; it thus contributes to a better understanding of neurotransmission and hormone signalling.

There are various ways to monitor exocytosis, and which one is adopted will depend on the specific interests of the experimenter and the type of excitable cell which is investigated. In essence, one will be detecting either the fusion of the vesicle with the surface membrane (membrane capacitance measurements; TIRF), or the material that comes out of the vesicle (amperometry), or the downstream effects of the released material on other surrounding cells (biosensors).

18.1.1 Amperometry

Amperometry is an electrochemical technique which allows the direct detection of readily oxidizable transmitters (e.g. adrenaline, noradrenaline, serotonin, dopamine) as they are released from a vesicle while, and after, it fuses with the plasma membrane. Illustrated in Figure 18.1, a thin carbon fibre electrode (CFE), which is insulated except at the active disc-shaped tip (≈ 5 μm diameter), is placed very close to a release site on the cell surface.

Proximity is crucial here – a distance of under ≈ 5 μm is required. The CFE is charged to an electrical potential (voltage) more positive than the transmitter's

Essential Guide to Reading Biomedical Papers: Recognising and Interpreting Best Practice, First Edition.
Edited by Phil Langton.
© 2013 by John Wiley & Sons, Ltd. Published 2013 by John Wiley & Sons, Ltd.

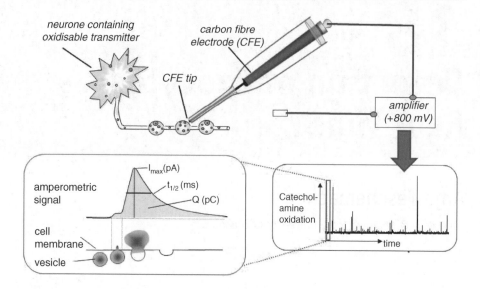

Figure 18.1   Diagrammatic representation of amperometric recording. Amperometric recording of oxidizable transmitters, e.g. catecholamine release from central neurones using a positively charged carbon fibre electrode (CFE), gives a spike shape consistent with diffusion of a transmitter bolus from a release site to the CFE. Spikes can be analysed for their amplitude (I_{max} in pA), time course ($t_{1/2}$ is time at 50 per cent I_{max} in msec), and their integral (Q in pCoulomb) which is a measure for quantal size. Adapted from (Kasparov S & Teschemacher AG (2008). Altered central catecholaminergic transmission and cardio-vascular disease. *Experimental Physiology* **93**, 725–740). © Physiological Society. *A full colour version of this figure appears in the colour plate section.*

redox potential (e.g. +650 mV for serotonin). Upon exocytosis of a vesicle, its transmitter content diffuses from the release site to the CFE tip and is instantly oxidized as it arrives. As oxidation involves the movement of electrons (from the material being oxidized), the rapid oxidation of the transmitter results in a spike-shaped electrical current signal.

The integral of the spike signal is generally assumed to be directly proportional to the quantity of transmitter molecules oxidized, and is thus a measure of quantal size.

18.1.2   Capacitance measurements

These are an indirect method of inferring vesicle exocytosis by monitoring changes in total cell surface membrane. The lipid bilayer structure of the plasma membrane makes it an excellent electrical capacitor (i.e. allows it to transiently store electric charge). The bigger the cell, the larger its plasma membrane and the higher its capacitance measured in Farads (F). When exocytosis occurs, vesicular membrane is added to the cell's membrane, thereby increasing total membrane capacitance (C_m) in proportion.

In practice, a cell is voltage-clamped using the whole-cell or perforated patch configuration of patch clamp (see Primer 11), and a square or sine wave voltage command is applied through the intracellular microelectrode. Any increase in capacitance will require a greater electrical charge movement from the electrode, resulting in a current that decays over a longer period of time before the now larger capacitance comes to equilibrium at the command voltage.

As good voltage-clamp conditions are critical, cells need to have an uncomplicated morphology (no long processes or gap junctions), so round cells in dissociated culture are ideal. Time resolution is usually in the ≈10 msec range. While resolution of most experimental systems is not high enough to pick up single fusion events of any other than very large vesicles, a handful of large, dense-core vesicles of ≈300 nm diameter are well within reach.

A further sophistication of C_m measurements is illustrated in Figure 18.2. '*Patch amperometry*' monitors the capacitance of the small patch of membrane under the tip of a microelectrode in on-cell patch configuration. This approach has allowed resolution of single fusion events by C_m recording, while simultaneously measuring oxidizable transmitter release through a CFE placed inside the microelectrode (Figure 18.2).

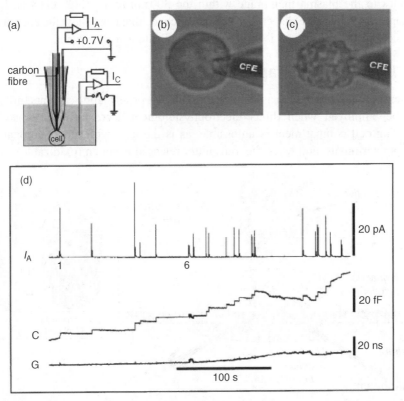

Figure 18.2   Patch amperometry. (a): Arrangement of a CFE inside a patch pipette. IA, Amperometric current; IC, sine wave current used to measure capacitance changes. (b, c): Chromaffin cell with attached patch pipette containing CFE at the beginning (b) and end (c) of the experiment. (d): Recording from this cell shows amperometric transients (top), associated capacitance steps (middle) and conductance trace (bottom), using a 20 kHz, 50 mV sine wave. C is capacitance trace, G is conductance trace. Reproduced, with permission, from Albillos, A., et al., The exocytotic event in chromaffin cells revealed by patch amperometry. *Nature*, **1997**. 389(6650): p. 509–512 © Nature.

18.1.3   Total internal reflection fluorescence (TIRF) microscopy

Also known as evanescent wave microscopy, TIRF is an optical method for visualizing single vesicles during approach and fusion with the plasma membrane. It requires fluorescently loaded or labelled vesicles in cells grown on glass cover slips, and a fluorescence microscope equipped for TIRF imaging.

A laser beam is directed through a prism from beneath the cell of interest at a 'critical angle'. The critical angle illumination is reflected at the glass-water interface of the plane, but not without also illuminating the first ≈300 nm of the water phase (in which the cells are growing). When fluorescent vesicles approach the cell membrane and enter this thin layer of illumination, the fluorescent marker molecules become excited and, by their emission light, become visible through the microscope. In this way, TIRF allows the experimenter to monitor the approach of vesicles to the plasma membrane within the 300 nm range (Figure 18.3). Rapid disappearance (termed 'dispersion') of fluorescence that can follow the approach of a vesicle to the membrane is interpreted to indicate a fusion event.

18.1.4   Biosensor approaches

Biosensor approaches, such as analysis of miniature postsynaptic currents (mPSCs), can be employed when direct electrophysiological access to the transmitter-releasing cell compartment is unfeasible, as is the case with most neuronal pre-synaptic terminals, and when the transmitter released is not an (oxidizable) amine.

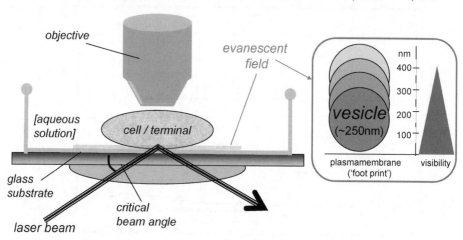

Figure 18.3   TIRF microscopy for visualization of fluorescently labelled vesicles. Vesicles are visualized as they approach the plasma membrane of a cultured cell or synaptic terminal facing the glass substrate. The laser beam, directed at a critical angle, illuminates an evanescent field about 300 nm into the cell (shaded red), within which vesicles are visible (inset), becoming brighter as they approach the plasma membrane. Courtesy of Dr. Anja Teschemacher. *A full colour version of this figure appears in the colour plate section.*

Biosensor approaches make use of the exquisite sensitivity and selectivity of ligand-gated ion channels tuned to their transmitter agonist.

In practice, the biosensor cell (e.g. a post-synaptic neurone) is recorded in voltage-clamp mode through an intracellular microelectrode (see Primer 10), or a whole-cell patch electrode (Primer 11). Pharmacological blockade of action potentials (with tetrodotoxin), and of ion channels irrelevant to the particular experiment (e.g. GABA$_A$ receptor block if glutamate release is to be studied), isolates mPSCs occurring in response to quantal release.

Cumulative histograms are plotted for the interval times between spontaneous mPSCs and for their amplitudes. Plots are then compared between experimental conditions, e.g. control versus drug. Changes in mPSC frequency are interpreted as a *pre*-synaptic effect, i.e. a modulation of *probability* of exocytosis, whereas changes in mPSC amplitude distribution point to *post*-synaptic action.

18.2 Common pitfalls in execution or interpretation and required controls

18.2.1 Amperometry

- There are numerous potential sources for false positive signals, e.g. electrical or mechanical interferences. Therefore, it is important that individual amperometric spikes are scrutinized and any excluded from analysis that do not fit with a diffusion-dependent signal shape. Reports should show examples of spikes at sufficiently high time resolution to demonstrate their shapes.

- False-positive signals may also arise from oxidizable substances other than the transmitter under investigation. For this reason, cyclic or stepwise variation of the CFE charging potential (i.e. driving voltage) needs to be carried out to confirm that events are caused by oxidation of a substance that requires a driving voltage that is consistent with the transmitter in question (see Figure 18.4 below).

- False-negative measurements are possible. Even under the best microscope, it can be very difficult to see thin cellular processes or to accurately identify likely release sites. Diffusion barriers that are practically invisible and release sites remote from the CFE can result in a failure to record release events. Absence of evidence (of release), however, does *not* equate to evidence of absence (of release).

- Quantal content may also simply be underestimated. Positioning of CFE close to the release site is crucial, because otherwise a fraction of transmitter will diffuse away from the CFE, escape oxidation and thus go undetected.

- CFE poisoning (i.e. loss of detector sensitivity over time) is common, predominantly due to oxidation product and debris deposits on the active

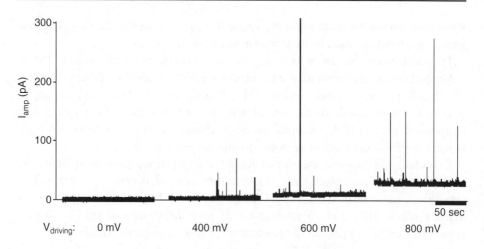

Figure 18.4   Norepinephrine release by amperometry. Noradrenaline release detected in organo-typic brain slice by amperometry at a range of driving voltages (0–800 mV). From Zohreh Chiti and Anja G. Teschemacher (2007). Exocytosis of norepinephrine at axon varicosities and neuronal cell bodies in the rat brain. The FASEB Journal vol. 21 no. 10 2540–2550

carbon surface. These deposits tend to reduce efficiency of oxidation and so attenuate signals. Reports should mention explicitly response stability over time and explain how this was monitored and managed. Typically, calibrations will be performed at intervals, and certainly at the start and end of an experiment, along with criteria for exclusion of data that is affected by electrode poisoning.

- Keep in mind that only oxidizable vesicle contents are detected. For example, most co-transmitters would not be detected.

18.2.2   Capacitance measurements

- C_m is an *indirect* measure of transmitter release, and critically depends on voltage-clamping the cell well. Therefore, C_m measurements are of limited use when investigating exocytosis in cells with complex morphology (e.g. neu-rones), because voltage-clamp conditions will be far from ideal.

- Changes in C_m cannot be measured during membrane conductance changes (e. g. ionic currents switching on or off). Therefore, if exocytosis is triggered via voltage-gated Ca^{2+} channels, measurements are interrupted for (at least) the duration of the stimulus, due to conductance changes. Stable conductance (G) recordings over time have to be confirmed.

- Experimental designs often include measures to inhibit (i.e. block) ion channels that are regarded as non-essential to the experiment. The possibility that drugs used for this purpose might influence the results in ways that are unappreciated should not be forgotten.

- C_m changes reflect the net result of both exo- and endocytotic events. Exocytosis may, therefore, be underestimated, depending on time course and magnitude of endocytosis occurring in parallel. It is generally assumed that endocytotic events are sufficiently delayed so that the two time courses do not significantly overlap.

- When recording in whole-cell patch clamp configuration, intracellular factors important for exocytosis (and endocytosis) may be dialysed out of the cell. Look out for reproducible C_m changes for the duration of the experiment (i.e. no run-down of exocytotic processes). This problem can be largely prevented by using the perforated patch configuration. Reports should address this issue in methods and/or discussion.

- Keep in mind that C_m reports all fusion events, irrespective of vesicle content. That means that there is no information on the transmitter type and quantity released, and vesicles could even be empty.

18.2.3   TIRF

- TIRF imaging is affected by the typical caveats of fluorescence microscopy approaches, such as photobleaching, etc (see Primers 8 and 16).

- Depending on the labelling approach used, intracellular structures other than exocytotic vesicles may become fluorescent, thereby decreasing the signal-to-noise ratio. Reports may demonstrate that fusion events (rapid dispersion of fluorescence) are stimulus- and/or Ca^{2+}-dependent. Such demonstrations will help to validate the claim that exocytosis is being visualized.

- A further potential limitation is that only release events at a base of the cell (termed 'footprint') can be detected. There are legitimate concerns over how representative of a physiological release-active membrane such areas are likely to be.

18.2.4   Biosensor – miniature postsynaptic current (mPSC) analysis

This is an indirect way of measuring pre-synaptic vesicle fusion that relies on recording and interpretation of postsynaptic responses. Interpretation of mPSC measurements depends critically on *assumptions* which may not hold true, depending on the cell type under investigation (see review by Vautrin & Barker, 2003). Some of these assumptions are as follows:

- The frequency of spontaneous vesicle fusion correlates tightly with the pre-synaptic release probability and is assumed to be a good predictor of transmitter release in response to action potentials. This appears to be the case in many scenarios, but examples have been described where action potential-evoked and spontaneous exocytosis in a cell are modulated via separate pathways.

- Quantal size is unaffected by the experimental challenge or modulator. Were quantal sizes to change, mPSC amplitudes would be expected to change, though not necessarily in strict proportion. The interpretation offered may focus entirely upon post-synaptic modulation (potentiation or depression) and ignore the possibility of altered pre-synaptic quantal size.

- The relative contributions of all the synapses signalling on to the postsynaptic biosensor remain constant overall. Neurones with an extensive receptive field (a dendritic tree) will sample transmitter released from many synapses originating from multiple pre-synaptic cells. As mPSCs are graded signals, they will become smaller as they propagate from their source, e.g. up on the dendritic tree, towards the recording electrode. This has two implications:

  ○ The mPSC size does not correlate well with the range of quantal sizes in complex (neuronal) biosensor cells.

  ○ If the experimental challenge or modulator affects pre-synaptic cells or terminals contacting part of the dendritic tree only, the resulting changes to the mPSC amplitude distribution may imply a postsynaptic modulation, while, in fact, release probability in these cells/terminals was affected.

18.3   Complementary and/or adjunct techniques

- Primer 10: Single electrode voltage-clamp

- Primer 11: Patch clamp

- Amperometry: Voltammetry (or, better, fast scan cyclic voltammetry); for some cell types such as chromaffin cells – ΔCm or patch amperometry

- Capacitance measurements: Intracellular Ca^{2+} measurements (see fluorescence microscopy); measurements of Ca^{2+} entry through voltage gated channels by voltage-clamp; release of caged Ca^{2+}.; TIRF; amperometry.

- TIRF: Voltage-clamp recordings and stimulation; measurements of Ca^{2+} entry and concentration; Cm measurements.

- Biosensor (mPSC analysis): Intracellular microelectrode recording, patch clamp, voltage-clamp.

Further reading and resources

Albillos, A., Dernick, G., Horstmann, H., Almers, W., Alvarez de Toledo, G. & Lindau, M. (1997). The exocytotic event in chromaffin cells revealed by patch amperometry. *Nature* **389**, 509–512.

Angleson, J.K. & Betz, W.J. (1997). Monitoring secretion in real time: capacitance, amperometry and fluorescence compared. *TINS* **20**(7), 281–287.

Chiti, Z. & Teschemacher, A.G. (2007). Exocytosis of norepinephrine at axon varicosities and neuronal cell bodies in the rat brain. *FASEB Journal* **21**(10), 2540–2550.

Kasparov, S. & Teschemacher, A.G. (2008). Altered central catecholaminergic transmission and cardiovascular disease. *Experimental Physiology* **93**, 725–740.

Teschemacher, A.G. (2005). Real-time measurements of noradrenaline release in periphery and central nervous system. *Autonomic Neuroscience: Basic and Clinical* **117**, 1–8.

Vautrin, J. & Barker, J.L. (2003). Pre-synaptic quantal plasticity: Katz's original hypothesis revisited. *Synapse* **47**, 184–199.

Wightman, R.M., Schroeder, T.J., Finnegan, J.M., Ciolkowski, E.L., & Pihel, K. (1995). Time course of release of catecholamines from individual vesicles during exocytosis at adrenal medullary cells. *Biophysical Journal* **68**, 383–390.

Zenisek, D., Steyer, J.A. & Almers, W. (2000). Transport, capture and exocytosis of single synaptic vesicles at active zones. *Nature* **406**, 849–854.

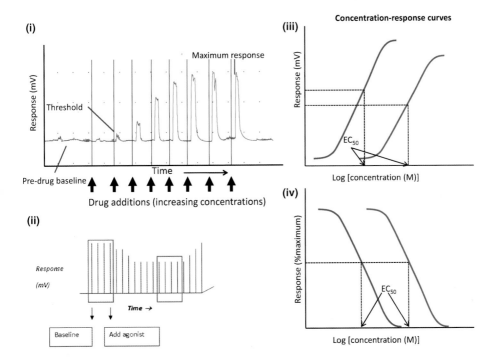

Figure 4.2 Concentration-response data illustrating agonist potency analysis in unstimulated and electrically stimulated tissue. The results shown in panels (i) and (ii) illustrate the type of output data which is obtained from an isolated tissue preparation in the stimulated and unstimulated set-up.

Panel (i) shows data for an incremental concentration response experiment recorded using a computer software program. Note how the two highest concentrations induce a similar level of response, indicating the maximum has been achieved. Increasing the concentration further is likely to desensitize the tissue.

Panel (ii) illustrates the type of data which can be recorded using the stimulated set-up. Electrical stimulation triggers the release of transmitter and contraction. For a given level of electrical stimulation, a stable response level is achieved. Drugs which inhibit the release of the transmitter can then be added and their effect observed as a reduction in the magnitude of the contraction.

Panel (iii) and (iv) show the results for full concentration response studies comparing two different agonists which induce contraction (panel iii) or relaxation (panel iv). Both types of data can be fit to a sigmoidal concentration-response curve and an EC50 value for each agonist obtained. Where more than one agonist is tested, as illustrated, their relative potency can be assessed. Drugs which are more potent sit to the left of the graph and achieve an EC50 at a lower concentration. Data may be plotted using an absolute value (panel iii) or normalised data (panel iv). Courtesy of Dr. Emma Robinson.

Essential Guide to Reading Biomedical Papers: Recognising and Interpreting Best Practice, First Edition.
Edited by Phil Langton.
© 2013 by John Wiley & Sons, Ltd. Published 2013 by John Wiley & Sons, Ltd.

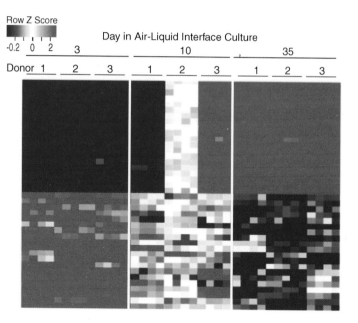

Figure 6.1 Variability in primary human bronchial epithelial cell gene expression *in vitro*. Messenger RNA was measured in quadruplicate using high throughput sequencing of ALI cultures from three different individuals at three time points as indicated. Forty of the most up- and down-regulated mRNAs are illustrated. Note that each person has a different time course and the intermediate time point is highly variable, which will introduce variability if not controlled for carefully. Also note variability in expression of down-regulated genes at the latest time point. Courtesy of Dr. Scott H. Randell.

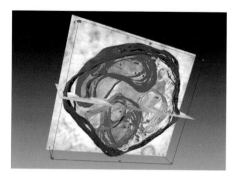

Figure 7.4 Example of a 3-D reconstruction and model of an endosome. After reconstruction of the volume, a specific slice can be analyzed. Specific features can be drawn into the reconstruction. Doing this on subsequent slices enables a 3-D model to be built. Courtesy of Dr. Paul Verkade.

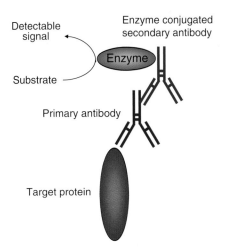

Figure 12.2 Schematic representation of primary and enzyme-conjugated secondary antibody reactions. After the primary antibody is bound to the target protein, a complex with enzyme-linked secondary antibody is formed. The enzyme (e.g. alkaline phosphatise, horseradish peroxidise), in the presence of a substrate, produces a detectable signal. This approach is commonly used in techniques such as ELISA, immunoblotting, immunocytochemistry and immunohistochemistry. Courtesy of Professor Elek Molnár.

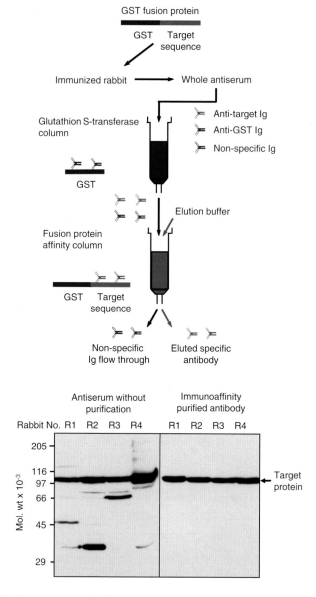

Figure 12.3 Purification of antibodies raised against a glutathione S-transferase (GST) fusion protein. The antiserum is first pre-adsorbed using an affinity column coupled with the GST protein. Any antibodies against GST are removed by this column. Specific antibodies to the target sequence flow into the second affinity column, which is coupled to a GST fusion protein containing amino acid sequences of the target epitope. Both columns allow the non-specific antibodies to flow through. Specific antibodies directed against the epitope region of the GST fusion protein are captured in the second column and eluted using a low pH buffer (0.1 M glycine-HCl, pH 2.5; red arrows). The elute is immediately neutralised by the addition of 1 M Tris-HCl (pH 8.0). Immunoblots illustrate reaction specificity of rabbit polyclonal antibodies before and after immunoaffinity chromatography purification. Full experimental details are available in Pickard *et al.* (2000). Courtesy of Professor Elek Molnár.

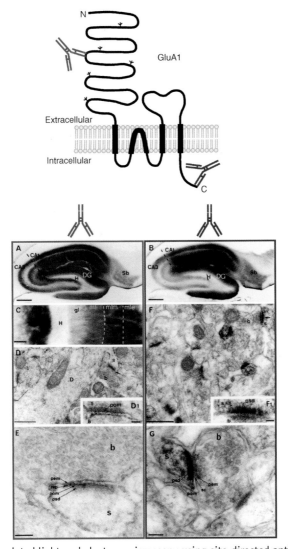

Figure 13.2 Correlated light and electron microscopy using site-directed anti-peptide primary antibodies in combination with horseradish peroxidise (HRP) enzyme-conjugated secondary antibody. Two rabbit polyclonal antibodies were generated against N-terminal (red) and C-terminal (blue) epitopes of the GluA1 AMPA receptor subunit protein using synthetic peptides (Molnar *et al.*, 1994). Immunolabelling with both antibodies indicates similar distribution for GluA1 in the rat hippocampus (A, B). Electron micrographs of GluA1 immunoreactivity (D-G) indicate that the HRP reaction end-product (rep) is concentrated at the extracellular face of the postsynaptic membrane (pom), demonstrating extracellular location for the N-terminal antibody binding site (E). In contrast, the reaction end-product obtained with the C-terminal antibody is located intracellularly at the postsynaptic membrane, on the postsynaptic density (psd), and not in the synaptic cleft (sc). Modified from Molnár *et al.* (1994). Scale bars: 0.5 mm (A and B), 50 μm (C), 0.5 μm (D and F), 0.05 μm (D1 and F1), 0.1 μm (E and G). Reproduced, with permission, from Molnár E, McIlhinney RAJ, Baude A, Nusser Z, Somogyi P (1994) Membrane topology of the GluR1 glutamate receptor subunit: Epitope mapping by site-directed anti-peptide antibodies. J Neurochem **63**:683–693. © John Wiley & Sons Ltd. Courtesy of Professor Elek Molnar.

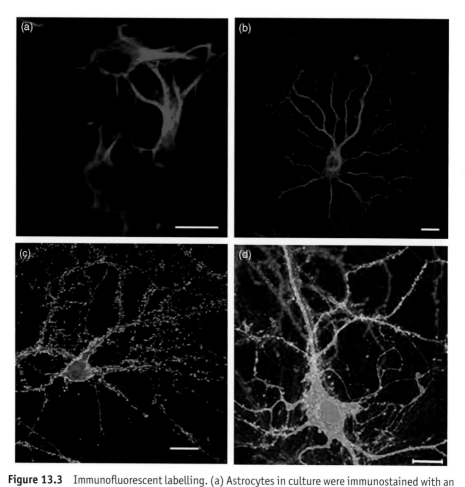

Figure 13.3 Immunofluorescent labelling. (a) Astrocytes in culture were immunostained with an antibody against glial fibrillary acidic protein (GFAP), which is a major component of intracellular skeleton of protein filaments in these cells (Luyt *et al.* 2003). PANEL (a) Reproduced, with permission, from Luyt K, Varadi A, Molnar E (2003) Functional metabotropic glutamate receptors are expressed in oligodendrocyte progenitor cells. J Neurochem 84:1452–1464. (b) Differentiated rat oligodendrocyte in primary culture, stained with an antibody to metabotropic glutamate receptor mGluR5 (Luyt *et al.* 2006). PANEL (b) Reproduced, with permission, from Luyt K, Varadi A, Durant CF, Molnar E (2006) Oligodendroglial metabotropic glutamate receptors are developmentally regulated and involved in the prevention of apoptosis. J Neurochem 99:641–656. (c) Co-localization of activated (autophosphorylated) calcium-calmodulin kinase II (CaMKII; green – an enzyme involved in long-term potentiation in the central nervous system synapses) and synaptotagmin (synaptic marker) in hippocampal neuronal cultures (Appleby *et al.* 2011). PANEL (c) Reproduced, with permission, from Appleby VJ, Corrêa SAL, Duckworth JK, Nash JE, Noël J, Fitzjohn SM, Collingridge GL, Molnár E (2011) LTP in hippocampal neurons is associated with a CaMKII-mediated increase in GluA1 surface expression. J Neurochem 116:530–543. (d) Co-localization of flag epitope-tagged truncated C-terminal cargo domain of myosin VI (green), GluA1 (red) and GluA1-4 (blue) AMPA receptor subunits using mouse, rabbit and guinea pig primary antibodies, respectively (Nash *et al.* 2010). Scale bars: 10 μm. PANEL (d) Reproduced, with permission, from Nash JE, Appleby VJ, Corrêa SAL, Wu H, Fitzjohn SM, Garner CC, Collingridge GL, Molnár E (2010) Disruption of the interaction between myosin VI and SAP97 is associated with a reduction in the number of AMPARs at hippocampal synapses. J Neurochem 112:677–690. © John Wiley & Sons Ltd.

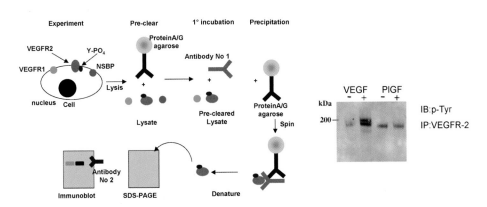

Figure 14.1 The steps involved in IP. A. After cell lysis, antigens that might bind non-specifically to protein A/G (non-specific binding protein NSBP), are removed by pre-clearing. Lysate is then incubated with primary (1°) antibody, the protein A/G agarose, spun, denatured and subjected to sodium dodecylsulphate-polyacrylamide gel electrophoresis (SDS-PAGE) and immunoblotting using a different primary antibody. B. An immuno- (Western) blot of a cell lysate of cells that have been exposed either to VEGF or PlGF, and then immunoprecipitated (IP) with an antibody to VEGFR-2. The protein has then been run on a polyacrylamide gel and immunoblotted (IB) with an antibody to phosphotyrosine (p-Tyr). It can be seen from the gel that VEGF, but not PlGF, can stimulate phosphorylation of one or two proteins around 200 kDa in these cells. VEGFR-2 is approximately 200 kDa, so this is interpreted as showing that VEGF, but not PlGF, can result in phosphorylation of VEGFR-2. Courtesy of Professor Dave Bates.

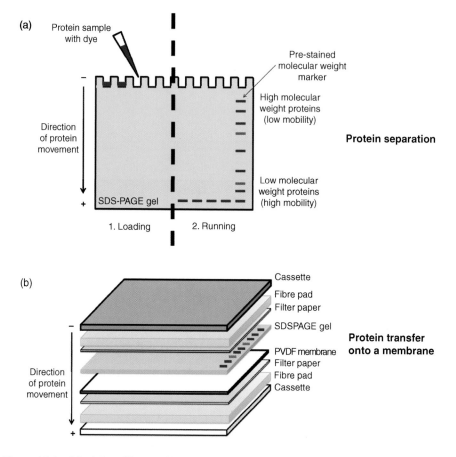

Figure 15.1 Principles of immunoblotting. (a) Protein separation: the protein sample is mixed with SDS and a dye and loaded on a SDS-PAGE gel. The SDS gives the protein (polypeptide) a uniform charge, which allows separation according to size. When a current is applied, the negatively charged proteins move in the direction of the positive (anode) pole. High molecular weight proteins move slowly and remain near the top of the gel, whereas low molecular weight proteins move faster and end up near the bottom of the gel. Pre-stained molecular weight markers (containing proteins of known molecular weight) are also run on the gel in order to estimate the apparent molecular weight of proteins detected in C. (b) Protein transfer onto a membrane: proteins are transferred from the gel onto a membrane by electro-elution. The negatively charged proteins move out of the gel towards the positive pole and bind to the membrane. During this procedure, the gel and membrane are in close contact and submerged in transfer buffer.

(Continued on next page)

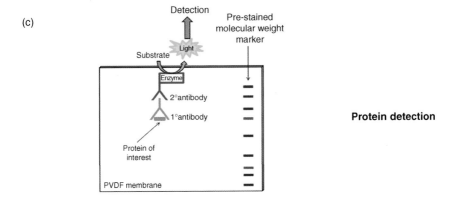

(c)

Detection

Pre-stained
molecular weight
marker

Light

Substrate

Enzyme

2°antibody

1°antibody

Protein of
interest

PVDF membrane

Protein detection

Figure 15.1 (*Continued*) Principles of immunoblotting. (c) Protein detection; the membrane is incubated with primary (1st) antibody against the protein of interest, followed by a enzyme labelled secondary (2nd) antibody that binds to the primary antibody. Enzymatic substrate conversion results in the production of light (chemiluminescence), which can be detected by X-ray film. Alternatively, fluorescently labelled secondary antibodies can be used. Courtesy of Dr. Ingeborg Hers.

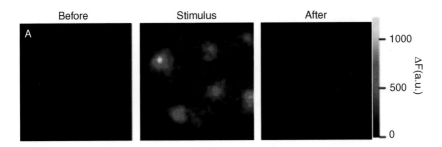

Before   Stimulus   After

A

1000

500

0

ΔF(a.u.)

Figure 17.1 Ca^{2+} microdomains mediated by Ca^{2+} influx at ribbon synapses in IHCs. Confocal images of Fluo-5N-filled IHCs were acquired before, during, and after stimulation by 200 ms depolarizations to -7 mV. Images averaged over runs and time frame are shown. (Scale bars: 2 μm.) Adapted from Thomas Frank, Darina Khimich Andreas Neef, and Tobias Moser. (2009) Mechanisms contributing to synaptic Ca^{2+} signals and their heterogeneity in hair cells PNAS 106 (11): 4483–4488. © National Academy of Science.

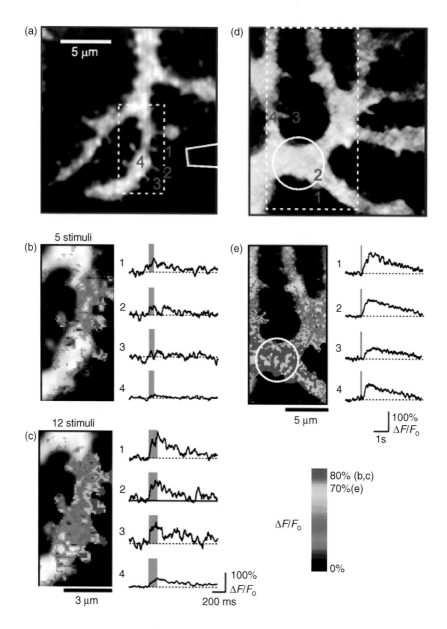

Figure 17.2 InsP3-mediated Ca^{2+} signals produced in Purkinje cell dendritic spines and shafts by parallel-fibre (PF) activity (left) and uncaging InsP3 (right). The area studied in (b, c) (outlined by dashed box) and the position of the PF stimulation pipette are shown. Numbers indicate locations from which Ca^{2+} signals were measured in (b, c). (b): Ca^{2+} signals produced by 5 stimuli (50 Hz) in CNQX. Traces represent Ca^{2+} signals in the numbered regions of (a). (c): Ca^{2+} signals produced by 12 stimuli (80 Hz). (d): Circle indicates position of the ultraviolet light spot (5 μm diameter) and numbers indicate locations from which Ca^{2+} signals were measured in (e). (e), Ca^{2+} signals produced by uncaging InsP3. Reproduced, with permission, from Elizabeth A. Finch & George J. Augustine (1998) Local calcium signalling by inositol-1,4,5-trisphosphate in Purkinje cell dendrites Nature 396, 753–756 © Nature.

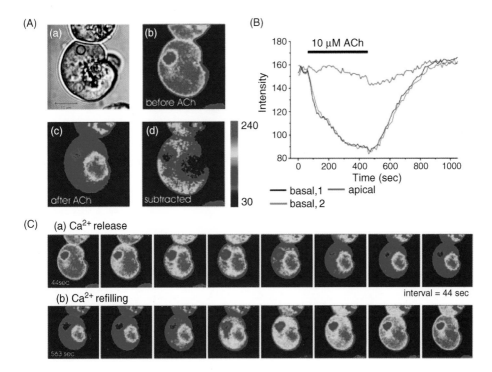

Figure 17.3 Time course of ACh-evoked Ca^{2+} release and the subsequent refilling of the ER. In a Mag-fluo 4-loaded cell, 10 μM ACh was applied. (A): (a) Transmitted light picture; (b and c) fluorescence intensity images before and after ACh; (d) subtracted image (b–c). (B): Time course of ACh-induced fluorescence intensity changes in the two basal regions as well as the apical regions marked in [A(a)]. (C): Time course of Ca^{2+} release following ACh stimulation and subsequent refilling after cessation of stimulation. Reproduced, with permission, from Fig 4 in Myoung Kyu Park, Ole H. Petersen and Alexei V. Tepikin (2000) The endoplasmic reticulum as one continuous Ca2+ pool: visualization of rapid Ca2+ movements and equilibration. The EMBO Journal 19, 5729–5739 © Nature.

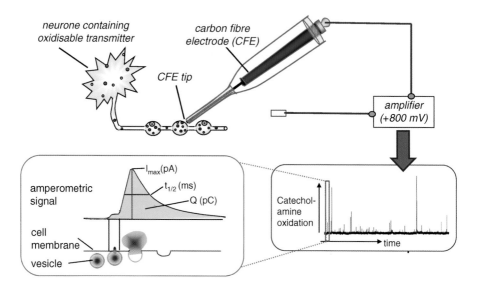

Figure 18.1 Diagrammatic representation of amperometric recording. Amperometric recording of oxidizable transmitters, e.g. catecholamine release from central neurones using a positively charged carbon fibre electrode (CFE), gives a spike shape consistent with diffusion of a transmitter bolus from a release site to the CFE. Spikes can be analysed for their amplitude (I_{max} in pA), time course ($t_{1/2}$ is time at 50 per cent I_{max} in msec), and their integral (Q in pCoulomb) which is a measure for quantal size. Adapted from (Kasparov S & Teschemacher AG (2008). Altered central catecholaminergic transmission and cardio-vascular disease. *Experimental Physiology* **93**, 725–740). © Physiological Society.

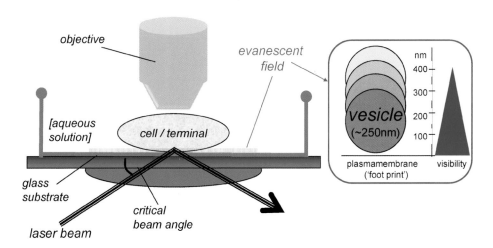

Figure 18.3 TIRF microscopy for visualization of fluorescently labelled vesicles. Vesicles are visualized as they approach the plasma membrane of a cultured cell or synaptic terminal facing the glass substrate. The laser beam, directed at a critical angle, illuminates an evanescent field about 300 nm into the cell (shaded red), within which vesicles are visible (inset), becoming brighter as they approach the plasma membrane. Courtesy of Dr. Anja Teschemacher.

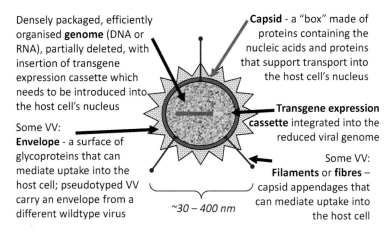

Figure 19.1 Salient features of VV which are derived from wild-type viruses and differ in detail between various types of VV available for research. Courtesy of Dr. Anja Teschemacher.

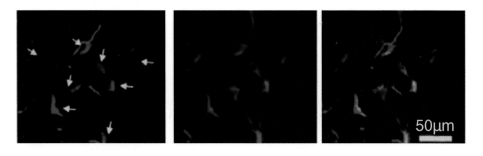

Figure 19.2 Independent confirmation of cell-specific transgene expression. Using immuno-fluorescence to confirm cell type-selective expression of a newly developed VV that uses a tryptophan hydroxylase 2-based shortened and enhanced promoter to express EGFP in central serotonergic neurones. Left: EGFP expression in green; middle: immunostaining for tryptophan hydroxylase 2; right: overlay. Reproduced, with permission, from Benzekhroufa K, Liu BH, Teschemacher AG, & Kasparov S (2009). Targeting central serotonergic neurons with lentiviral vectors based on a transcriptional amplification strategy. Gene Therapy 16, 681–688. © Nature.

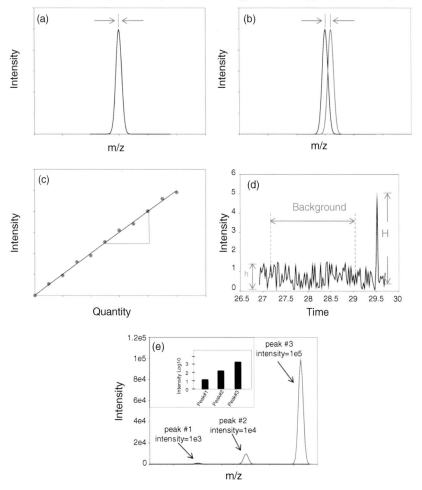

Figure 24.1 Visual representation of the different major characteristics that distinguish the different mass spectrometers. (a): the mass accuracy. (b): resolving power. (c and d): Sensitivity and limit of detection (LOD). (e): dynamic ranges. Courtesy of Dr. Thierry Lebihan.

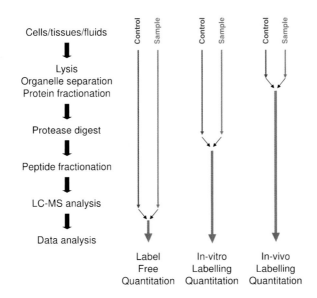

	Mass accuracy	Resolving power	Dynamic range	Sensitivity	Quantitation
qTOF	Good	Good	Medium	Good	Good
triple quadrupole	Medium	Low	High	High	Good
Ion trap	Low	Low	Low	Good	Poor
FT-ICR	Excellent	High	Medium	Medium	Medium
Orbitrap	Excellent	High	Medium	Medium	Medium

Figure 24.2 Main types of mass spectrometers and their characteristics. Courtesy of Dr. Thierry Lebihan.

Cells/tissues/fluids

Lysis
Organelle separation
Protein fractionation

Protease digest

Peptide fractionation

LC-MS analysis

Data analysis

| Control | Sample | Control | Sample | Control | Sample |

Label
Free
Quantitation

In-vitro
Labelling
Quantitation

In-vivo
Labelling
Quantitation

Figure 24.3 Typical quantitative mass spectrometry workflows. The left part is a generic process presented for a bottom up proteomics analysis. Red and Blue lines indicate when the two samples are normally combined together. The longer the process is run in parallel prior mixing the sample, the more technical variations are introduced, which can affect both samples independently. Courtesy of Dr. Thierry Lebihan.

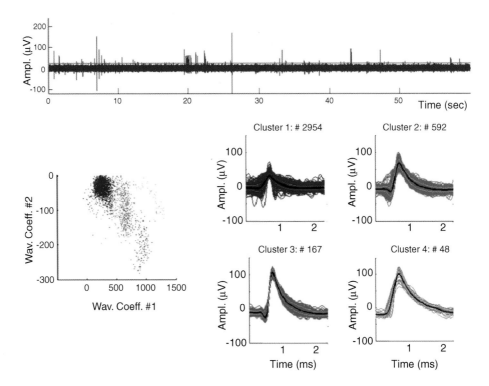

Figure 29.3 Example to illustrate how spike sorting can be used to discriminate individual neurones within a multi-unit trace. The top trace shows an extracellular multi-unit recording, in this particular case obtained with a microwire electrode positioned in the hippocampus. The red line above the trace indicates the threshold used for signal detection (i.e. all signals below this line are ignored). Waveforms above the line were analyzed to generate two 'wave coefficients' for each spike, each coefficient being based on particular aspects of the shape of the action potential. The lower graph on the extreme left shows a plot whereby each action potential (represented by a dot) is positioned according to its two coefficients. Note how this results in four distinct clusters of spikes (indicated by the different colours), which represent four distinct waveforms that have been discriminated. The averaged waveforms (in black) and superimposed individual waveforms (coloured) for these four spike clusters are shown in the lower right of the figure. Numbers above the waveforms refer to the number of spikes that have been included. Reproduced, with permission, from Quian Quiroga R, Reddy L, Koch C, Fried I (2007). Decoding visual inputs from multiple neurons in the human temporal lobe. Journal of Neurophysiology 98, 1997–2007. © American Physiological Society.

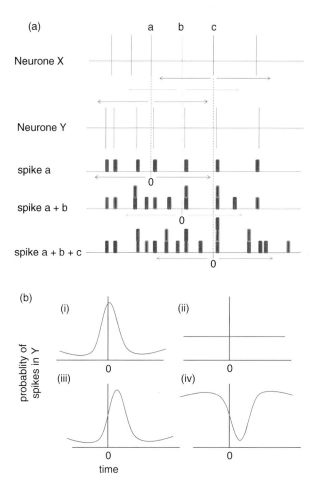

Figure 31.2 Aspects of analysis of spike-triggered averaging – the cross-correlogram (a) Construction of a cross-correlogram. Spike trains of two neurones recorded simultaneously. Time of occurrence of each spike in X is taken as time zero, and the time of spikes in Y are plotted on a histogram in relation to each time zero marker. Construction of a peri-event histogram is shown in relation to time of three spikes in X (spikes a, b and c). In this example, there is an increase in frequency of occurrence of spikes in Y just after a spike occurs in X. This is consistent with a monosynaptic excitatory connection between X and Y. (b) Examples of different shapes of cross-correlogram. (i) X and Y are synchronously active; (ii) activity of X and Y are not correlated; (iii) an increase in probability of spikes in Y occurs after activity in X, consistent with X exciting Y; (iv) a reduction in probability of spikes in Y occurs after activity in X, consistent with X inhibiting Y. Time zero = occurrence of spikes in X. Courtesy of Professor Richard Apps.

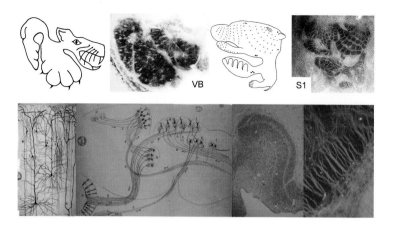

Figure 32.1 Visualizing neurons in the CNS. **Upper left panels**: Monoclonal antibody Cat-301 is directed against the cell-surface proteoglycan chondroitin sulphate of neurons and targets specific structures in the CNS. The section on the right shows an example of Cat-301 immunoreactivity in the cat. The section was taken in the oblique coronal plane through the first-order somatosensory nucleus of the thalamus, the ventrobasal complex (VB). Cat-301 stains neurons in the entire VB giving this structure a compartmentalized appearance. Rostral is at the top and medial is to the right. Based on receptive field mapping studies, the drawing on the left is a caricature of the cat body surface and reflects the compartments of the Cat-301 neuronal stain shown on the right. Figure 5 within Crabtree JW. (1999). Intrathalamic sensory connections mediated by the thalamic reticular nucleus. Cell Mol Life Sci. 1999 Nov 15;56(7–8):683–700. © Springer Publishing. **Upper right panels**: Example of staining for the mitochondrial enzyme succinic dehydrogenase in the rat. The section on the right was taken tangential to the cerebral cortex through layer 4 of the flattened primary somatosensory cortex (S1). Lateral is at the top and rostral is to the left. Based on receptive field mapping studies, the drawing on the left is a caricature of the rat body surface and reflects the morphological pattern shown on the right. Reproduced, with permission, from Dawson DR, Killackey HP. (1987). The organization and mutability of the forepaw and hindpaw representations in the somatosensory cortex of the neonatal rat. J Comp Neurol. 1987 Feb 8;256(2):246–56. © John Wiley & Sons Ltd. **Lower left panels**: Drawings by Ramón y Cajal based on observations from Golgi-stained preparations. The drawing on the left shows 'First, second and third layers of the precentral of the cerebrum of a child of one month'. Projected onto the coronal plane, details of the cerebral cortex including dendritic spines are shown. The drawing on the right shows 'Scheme of the afferent and efferent pathways of the optic centers'. Projected onto the sagittal plane, connections of the retina (A) to the dorsal lateral geniculate nucleus (dLGN; D) and the superior colliculus (E) and the projection from the dLGN to the visual cortex (G) are shown. Drawings reproduced with the permission of the Inheritors of Santiago Ramón y Cajal ©. **Near right lower panel**: Nissl stain of the first-order visual nucleus of the thalamus, the dorsal lateral geniculate nucleus, of a cat (Crabtree, unpublished). Two laminae of neurons are clearly seen in the sagittal plane, an outer A lamina receiving inputs from the contralateral retina and an inner A1 lamina receiving inputs from the ipsilateral retina. Dorsal is at the top and rostral is to the right. **Far right lower panel.** Example of visualizing transparent brain tissue by using only the properties of light, that is, without using stains (Crabtree, unpublished). The section was taken in the horizontal plane through the internal capsule at the level of the thalamus in the cat. Two light sources were shone onto the section – one from above and one from below – producing contrast interference patterns and revealing large stringy bundles of axons (fasciculi) in the internal capsule (lower left to upper right) that de-fasciculate (upper left) as the axons approach the cerebral cortex. It is important to note that the techniques used to trace pathways in the CNS are generally too fine-grained to reveal large organizational features in the CNS, such as the maps shown in the upper panels. Courtesy of Dr. John Crabtree.

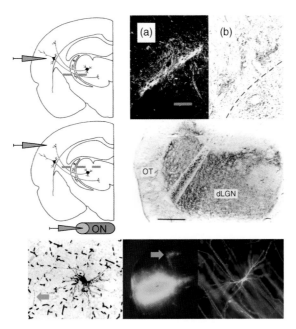

Figure 32.2   (*Continued on next page*)

Figure 32.2 (*Continued*) Pathway tracing in the CNS. **Top panels**: Example of anterograde tracing using a tritiated amino acid in the cat (from Crabtree 1992). The schema on the left shows the experimental setup projected onto the coronal plane. Injection of a tracer was made in layer 6 of the cerebral cortex, the neurons of which project to the thalamic reticular nucleus (R) and thalamic relay nuclei. The dashed line shows where sections were taken in the horizontal plane. (a, b): sections through R. Rostral is at the top and medial is to the right. (a): section showing silver grains following an injection of [3H]proline made in the cortex (dark-field optics). The calibration bar is 200 μm. (b): same section and field of view (Nissl stain). The dashed line indicates the anterolateral border of a thalamic relay nucleus. (left) Reproduced, with permission, from Crabtree JW. (1992). The Somatotopic Organization Within the Cat's Thalamic Reticular Nucleus. Eur J Neurosci. 4(12): 1352–1361. © Wiley. (right) Courtesy of Dr. John Crabtree. **Middle panels**: Examples of anterograde and retrograde tracing using the enzyme horseradish peroxidase (HRP) in the rabbit (Crabtree, unpublished). The schema on the left shows the experimental setup projected onto the coronal plane. Injections of HRP were made in the contralateral optic nerve (ON) and in layer 4 of visual cortex, which contains terminals of the axonal projections from the dorsal lateral geniculate nucleus (dLGN) of the thalamus. The dashed line shows where sections were taken in the horizontal plane. A section showing the resultant HRP labelling is shown on the right. Almost the entire dLGN is filled with anterograde labelling of terminals where HRP was transported via the axons of the ON and optic tract (OT). A small label-free region representing input from the ipsilateral ON can be seen on the far right. The dLGN also contains a column of retrograde labelling of cell bodies (insert) where HRP was transported via the axons of dLGN relay cells. Rostral is at the top and medial is to the right. The calibration bar is 500 μm. **Lower left panel**: Example of a biocytin-filled neuron in a thalamic relay nucleus of the rat (from Crabtree and Isaac 2002). Intracellular injection of biocytin gives neurons a Golgi-like appearance. This allows visualization not only of the trajectory and extent of single axons and their branches but also of morphological features such as the size and shape of the neuronal soma and dendritic tree. The arrow points to the axon of the biocytin-filled neuron. **Lower right panels**: Example of retrograde tracing using the fluorescent carbocyanine dye DiI in the rabbit (from Crabtree 1999). Two sections taken in the coronal plane through the thalamus are shown. On the left, labelling is shown after a small crystal of DiI was placed in the dorsal part of the thalamus (low magnification). The arrow points to a cluster of labelled neurons in the thalamic reticular nucleus. On the right, a higher magnification of one such labelled neuron is shown against a background of labelled axons. Unlike other anterograde and/or retrograde tracers, the carbocyanine dyes do not use the neuronal axoplasmic transport systems; rather, they diffuse along the extracellular surface of a neuron and its processes. It is important to note that the techniques used to trace pathways in the CNS are generally too fine-grained to reveal large organizational features in the CNS, such as the maps shown in the upper panels of Figure 32.1. Reproduced, with permission, from Crabtree JW. (1999). Intrathalamic sensory connections mediated by the thalamic reticular nucleus. Cell Mol Life Sci. 1999 Nov 15;56(7–8):683–700. © Springer.

19
Viral Vector Transgenesis

Anja Teschemacher
Physiology & Pharmacology, University of Bristol, UK

19.1   Basic 'how-to-do' and 'why-do' section

Viral vector transgenesis is an efficient method to introduce foreign genes – so-called transgenes – into living cells and to induce the cells to express these transgenes.

19.1.1   Viral vectors

Viral vectors (VV) are recombinant viruses, that is to say they are viruses modified by molecular cloning. By their nature, viruses are professional intracellular parasites. Their genomes are very condensed and lack the information for energy production and protein synthesis; therefore, they replicate literally by hijacking the protein synthesis machinery of their host cells. Viruses propagate by self-assembly of components that they induce their host cells to build, rather than by division. And, most importantly for the purpose of this primer, they have evolved highly efficient mechanisms for introducing their genes into host cells. This latter ability is retained in VV and harnessed for use in experiments that require the expression of non-native genes.

VV vectors are essentially disabled viruses, because significant parts of their genome have been deleted. As a result, VV are unable to replicate and therefore to cause a disease, making them safe to use in the laboratory. They also do not interfere with cells' normal functions, so that cell physiology can be studied. Deleting part of the native viral genome helps to clear space within the VV capsid that is necessary to accommodate the transgene 'cargo' which they are designed to deliver.

A major advantage of VV is that they are more efficient and versatile tools for gene delivery than any currently known chemical reagents over a wide range of experimental life sciences applications. They can be applied *in vitro* or *in vivo* and,

Essential Guide to Reading Biomedical Papers: Recognising and Interpreting Best Practice, First Edition.
Edited by Phil Langton.
© 2013 by John Wiley & Sons, Ltd. Published 2013 by John Wiley & Sons, Ltd.

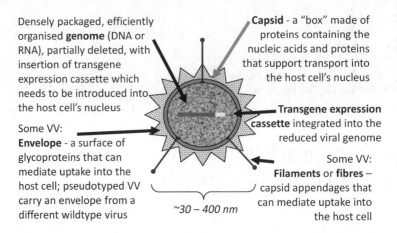

Densely packaged, efficiently organised **genome** (DNA or RNA), partially deleted, with insertion of transgene expression cassette which needs to be introduced into the host cell's nucleus

Some VV: **Envelope** - a surface of glycoproteins that can mediate uptake into the host cell; pseudotyped VV carry an envelope from a different wildtype virus

~30 – 400 nm

Capsid - a "box" made of proteins containing the nucleic acids and proteins that support transport into the host cell's nucleus

Transgene expression cassette integrated into the reduced viral genome

Some VV: **Filaments** or **fibres** – capsid appendages that can mediate uptake into the host cell

Figure 19.1 Salient features of VV which are derived from wild-type viruses and differ in detail between various types of VV available for research. Courtesy of Dr. Anja Teschemacher. *A full colour version of this figure appears in the colour plate section.*

to some extent, they work even across species. Moreover, some of the experience gained from experimentation is applicable to human gene therapy applications.

There is an increasing choice of VV systems available, and which one is adopted will depend on the research question, and the cell type which is investigated. Some key features that distinguish VV types (see Figure 19.1) are:

- Their transduction spectrum, characterized by their;

  ○ target cell selectivity, i.e. which cell types they can transfect;

  ○ transduction efficiency, i.e. how well they do it;

  ○ whether they enter from any part of the cell (e.g. in neurones this may be axonal terminals, or the cell body, or both).

- The most suitable route of application, e.g. whether they are most suitable for *in vitro* or *in vivo* experimentation.

- The stability of transgene expression – whether it declines after a few days or is robust for months.

- Whether the VV 'genome' integrates into the host cell's genome or remains outside of it – as so called 'episomal' genome.

- Their immunogenicity – some VV elicit more prominent immune responses than others, so are eliminated by the body's immune system more rapidly and can lead to local inflammatory or even shock reactions.

- Their capacity for 'cargo', i.e. what size of transgene they can package and deliver.

- Their production cost and effort – although generally much less costly and involved than production of transgenic animals, some VV are easier and cheaper to construct and amplify than others.

- The biohazard implications – since VV are genetically modified organisms, and derived from potential pathogens, their use is governed by strict biosafety regulations. Because of the viruses they have been derived from, some VV require handling under more stringent conditions than others.

19.1.2   Transgenesis

A transgene contains the coding sequence of the protein to be expressed. Various transgenes may be delivered for research purposes. For example, viral vectors are often used to express fluorescent proteins, permitting the experimenter to identify cells of interest and carry out single-cell experiments on them under a fluorescence microscope. Some fluorescent proteins can be used as reporters of cellular activity and concentration of ions (see Primer 17) or other cellular messengers. Other transgenes may change cellular activity (e.g. by gain or loss of function), and this helps in understanding the roles these cells of interest play within homoeostatic systems. Some VV contain transgenes that achieve both goals (visualization and functional change) simultaneously.

A very powerful approach is to direct transgene expression to selected tissues/ organs or even cell types. Transduction of a particular area can be achieved by topical administration, e.g. through microinjection, of the VV into the area of interest. Specific cell types can be targeted using cell-specific promoters for transgene expression. Each cell in the body produces its individual set of proteins, because each cell has a specific blend of active transcription factors that determine which endogenous proteins are expressed. If promoters are employed which are activated by the transcription factor complement present in a cell type of interest, transgene expression can be restricted to this cell type.

19.2   Common pitfalls in execution or interpretation & required controls

- Every VV is only as good as the combination of its ability to transduce cells of interest and the promoter that drives its transgene expression. Promoter activity is influenced by multiple factors and determines strength and stability of protein expression:

  o Promoter activity may vary, and some promoters are active in a wide range of cells while others only work in some cell types.

  o Some promoters take longer than others to build up sufficient transgene expression levels, while others have the tendency to become silenced over time.

The bottom line is simple. The use of any VV in a novel context (e.g. in a not previously studied tissue/area/cell type/age group/species) requires that the authors have successfully validated their approach by confirming successful transgene expression in the cells of interest by independent means such as immunohistochemistry (see figure 19.2).

- Strength of transgene expression also depends on how many VV particles transduce each cell; two copies of transgene will typically result in higher protein levels than a single copy, etc. It is generally believed that, in most situations *in vivo,* multiple VV transduce each individual cell; this is referred to as 'multiplicity of infection' (MOI) and may be as high as hundreds of VV genomes. Therefore, the titre of the VV batch (usually declared as 'transducing units per ml' or TU/ml), and the volume which is applied, are important determinants of transcription levels. Unfortunately, determining VV titres is not straightforward, and use of different approaches between researchers can make it difficult to compare between studies.

- Some VV types will cause residual expression of wild-type virus proteins and this can lead to immune responses if they are applied *in vivo*. Depending on the target tissues, immune responses can lead to degradation of VV and, hence, decreased transgene expression – or, for topical administration, to local tissue damage, loss of cells of interest and, again, decreased transgene expression.

- There are a few types of VV in experimental use (e.g. pseudorabies and Semliki forest virus-based 'vectors') which are, in fact, only partially disabled wild-type viruses. These actually do multiply in their host cells, albeit more slowly than their wild-type ancestors, and they eventually kill the cells. Experiments can be carried out within a limited time window between sufficient transgene

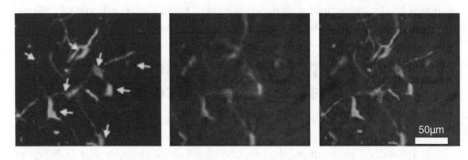

Figure 19.2 Independent confirmation of cell-specific transgene expression. Using immunofluor-escence to confirm cell type-selective expression of a newly developed VV that uses a tryptophan hydroxylase 2-based shortened and enhanced promoter to express EGFP in central serotonergic neurones.Left: EGFP expression in green; middle: immunostaining for tryptophan hydroxylase 2; right: overlay. Reproduced, with permission, from Benzekhroufa K, Liu BH, Teschemacher AG, & Kasparov S (2009). Targeting central serotonergic neurons with lentiviral vectors based on a transcriptional amplification strategy. Gene Therapy 16, 681–688 © Nature. *A full colour version of this figure appears in the colour plate section.*

expression levels and cell death – a potential caveat for interpretation of the physiological relevance of experimental results.

19.3   Complementary and/or adjunct techniques

Molecular cloning.
    Primer 6: Cell culture.
    Primer 8: Fluorescence microscopy.
    Primer 13: Immunohistochemistry.
    Primer 14: Immunoprecipitation.
    Primer 15: Immunoblotting.
    Primer 16: GFP.
    Primer 26: Transgenic animals.

Further reading and resources

Benzekhroufa, K., Liu, B.H., Teschemacher, A.G. & Kasparov, S. (2009). Targeting central serotonergic neurons with lentiviral vectors based on a transcriptional amplification strategy. *Gene Therapy* **16**, 681–688.

Davidson, B.L. & Breakefield, X.O. (2003). Viral vectors for gene delivery to the nervous system. (Review: 176 refs). *Nature Reviews Neuroscience* **4**, 353–364.

Duale, H., Kasparov, S., Paton, J.F.R. & Teschemacher, A.G. (2005). Differences in transductional tropism of adenoviral and lentiviral vectors in the rat brainstem. *Experimental Physiology* **90**(1), 71–78.

Lonergan, T., Teschemacher, A.G., Hwang, D.Y., Kim, K-S., Pickering, A.E. & Kasparov, S. (2005). Targeting brain stem centers of cardiovascular control using adenoviral vectors: Impact of promoters on transgene expression. *Physiological Genomics* **20**, 165–172.

Teschemacher, A.G., Paton, J.F. & Kasparov, S. (2005). Imaging living central neurones using viral gene transfer. *Advanced Drug Delivery Reviews* **57**, 79–93.

Teschemacher, A.G., Wang, S., Lonergan, T., Duale, H., Waki, H., Paton, J.F. & Kasparov, S. (2005c). Targeting specific neuronal populations in the rat brainstem using adeno- and lentiviral vectors: applications for imaging and studies of cell function. *Experimental Physiology* **90**, 61–69.

20
Polymerase Chain Reaction (PCR) and Reverse Transcription (RT)-PCR

Lucy F. Donaldson

Physiology & Pharmacology, University of Bristol, UK

20.1 Basic 'how-to-do' and 'why-do' section

In essence a method used to detect the presence of DNA in a sample, PCR has become one of the most important and widely used techniques in molecular biology. PCR is a technique by which minute amounts of DNA can be amplified from almost any sample to enable visualization, cloning, sequencing and further manipulation of DNA – manipulations that, until relatively recently, would have been impossible. The technique has been further developed over the last 20 years, so that now the basic PCR technique underpins more advanced molecular approaches such as real-time quantitation of DNA and RNA levels, cycle sequencing, and cloning and detection of mutation and polymorphisms.

Reverse transcription PCR (RT-PCR) is a modified PCR technique in which the starting point is a messenger RNA molecule (mRNA), usually isolated from a tissue sample or cell lysate. This mRNA is converted to DNA using a viral enzyme that uses RNA as a template to generate complementary DNA (cDNA) – reverse transcriptase. This has enabled improvement in the sensitivity of RNA analysis over and above the techniques used previously (Northern blots, slot blots, RNase protection assays).

Further developments in PCR technology have enabled detection of DNA amplification in real time (real time-PCR, confusingly also widely known as RT-PCR), reducing the necessity to run DNA samples on gels and increasing sensitivity. RNA amplification techniques (which are not covered here) have

Essential Guide to Reading Biomedical Papers: Recognising and Interpreting Best Practice, First Edition.
Edited by Phil Langton.
© 2013 by John Wiley & Sons, Ltd. Published 2013 by John Wiley & Sons, Ltd.

facilitated quantitation of minute amounts of RNA, and this has led to transcript analysis in single cells. PCR is now also used *in situ* to amplify RNA and DNA in cells in tissue sections. For further PCR applications, see Larrick & Siebert (1995).

PCR, RT-PCR and real time-PCR are now largely automated processes and can be done using high-throughput approaches, in which thousands of samples can be analyzed simultaneously.

20.1.1   Brief general method

Step 1   The target for amplification is isolated. This can be genomic DNA, viral DNA or mRNA. Tissue is homogenized or chemically digested, and nucleic acid is isolated by extraction with organic solvents (phenol/chloroform).

Step 2   The DNA or RNA to be used as a template for amplification is usually analyzed to determine DNA concentration and purity. This is often done by spectrophotometric analysis, but more sensitive techniques are now also available to allow quantification of very small amounts of nucleic acid.

RT-PCR   RNA is reverse transcribed into (complementary) DNA using Reverse Transcriptase. The commercially available forms of this enzyme have both DNA polymerase and RNase H activity. The former (polymerase) is the enzymatic activity that makes the DNA template. RNase H activity degrades the starting RNA molecules, making sure that they are no longer present to compete with DNA in further PCR reactions.

Step 3   The DNA, either genomic DNA, viral DNA or reverse transcribed cDNA, is included in a reaction mixture containing:

- buffer to maintain pH and ionic concentrations at ideal for the DNA polymerase enzyme;

- deoxynucleotides – the 'building blocks' of DNA, dATP, dCTP, dGTP, dTTP;

- primers[1] complementary to the sequence that will be amplified;

- DNA polymerase.

Step 4   The reaction mixture is cycled through, usually, three different temperatures:

---

[1] **Primer design.** Primers are short (15–30 bases) single-stranded pieces of DNA, designed to be complementary to the strands of the DNA to be amplified. There are two primers, one at each end of the bit of DNA to be amplified, each of which is complementary to one strand of the double-stranded target DNA molecule. These primers will base pair, 1 on each strand, at opposite ends of the sequence to be amplified, onto the target DNA and act as starting points for the DNA polymerase enzyme.

1. Separation: DNA strands are denatured (separated in single strands) by heating to ≈95°C.

2. Annealing: a short period at a temperature just below the melting point of the primers (the temperature at which the primer/target DNA duplex 'melts', i.e. comes apart), optimized to allow the primers to base pair to the target in the correct position.

3. Elongation: the mixture is held at 72°C, the temperature optimum of the DNA polymerase enzyme. This is the stage at which new DNA is synthesized.

The mixture is cycled through these temperatures, usually between 20 and 40 cycles, by which point the DNA has been amplified to an extent that it can be visualized by gel electrophoresis or directly measured by spectrophotometry (single copy becomes 1×10^{12} copies after 40 cycles).

DNA amplification by PCR approximates to a logistic (i.e. sigmoidal) curve. The initial amplification is exponential as, in the first cycle, two DNA strands are copied to give four; in the second cycle, four DNA strands become eight, and so on. As the reaction proceeds, amplification is linear (i.e. directly proportional to cycle number) and then, at some point, one of the constituents of the reaction will be exhausted and amplification will slow and ultimately stop in the plateau phase (see Figure 20.1).

Real time-quantitative (q)PCR allows highly sensitive detection of RNA transcripts and DNA using the incorporation of a fluorescent moiety in real time. The

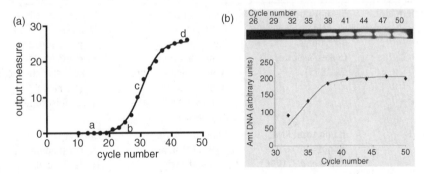

Figure 20.1   The relationship between cycle number and product for the PCR reaction. (a) The PCR reaction approximates to a logistic function, with no amplification or 'ground phase': (a), exponential phase (b), linear (c) and plateau (d) phases. The best results for any quantitative analysis, either by RT-PCR or real time-PCR, will be obtained from amplification cycles that are within the linear amplification phase. The output measure can be DNA-band optical density on a gel, spectrophotometric quantification or fluorescence signal for real time-PCR. (b) Illustration of the amplification of DNA as related to PCR cycle number. Top panel shows the DNA products from an ethidium bromide-stained gel, and the graph shows the relationship between PCR cycle number and DNA product, as determined by measurement of the optical density of the bands on the agarose gel. (left) Reproduced, with permission, from Sue Oldfield, Blair D Grubb, Lucy F Donaldson (2001). Identification of a prostaglandin E2 receptor splice variant and its expression in rat tissues. Prostaglandins & Other Lipid Mediators. 63: (4), 165–173 © Elsevier (right) Courtesy of Dr. Lucy Donaldson.

addition of probes that only fluoresce when the primer and probe are bound to the double-stranded PCR product greatly improves specificity. As the DNA amplification can be monitored as it occurs, the linear phase of the amplification can be identified in real time and, when an internal standard (a known amount of starting RNA, or DNA) is included, this method can be absolutely quantitative. In reality, real time-PCR is often used, as with classical PCR, in a semi-quantitative manner, to determine changes in amounts of transcript relative to a control condition, rather than absolute amounts of starting RNA.

20.2   Required controls

PCR, RT-PCR and, especially, real-time PCR, are very sensitive techniques in comparison to older, hybridization techniques such as RNase protection assays (see Table 20.1) and are easily confounded by contamination. Contamination with either cDNA or genomic DNA can lead to false positive results. cDNA for common targets in all PCR and RT-PCR experiments should have controls for general contamination (e.g. inclusion of tRNA rather than the DNA sample) and for genomic DNA (inclusion of a sample of non-reverse transcribed RNA).

Table 20.1 Comparison of relative sensitivities, advantages and disadvantages of different techniques for studying RNA transcript expression (Cline *et al.*, 1996; Becker-Andre, 1995).

Technique	Sensitivity	Advantages	Disadvantages
Northern blots	5–10 µg polyA+ RNA 10^6 copies (Sykes & Morley, 1995).	Semi-quantitative. mRNA size. Shows any mRNA degradation. Shows splice variants Single blot can be reused multiple times.	Not very sensitive. Requires complex procedure and equipment. Use of toxic chemicals (formaldehyde) and radioisotopes.
RT-PCR	<1 ng total RNA. Single cell. (Cline *et al.*, 1996).	Quantitative (e.g. PATTY) semi-quantitative. Rapid. Sensitive. High throughput – analysis of multiple transcripts.	Subject to contamination – false positives. May not identify splice variants. Absolute quantitation requires complex protocols.
Real time-PCR	Can detect single copy target cDNA (Cline *et al.*, 1996).	Quantitative/Semi-quantitative. Highly specific (depends on method). Highly sensitive. Rapid.	High cost. May be too sensitive – subject to false positives. No information on unidentified splice variants.

A water control (i.e. with no nucleic acid) is also sometimes included. This control will determine whether any of the other reaction constituents is contaminated with nucleic acid. One of the best controls for genomic DNA contamination is to design primers that span introns in the genomic sequence. Should genomic DNA be present, when amplified with intron-spanning primers, it will be amplified to give a longer product than the cDNA target, due to the inclusion of an intronic sequence. 'No primer' reactions can also be used as controls. These are all examples of *negative controls* (Taylor, 1991).

Positive controls should also be included, to ensure that a lack of PCR product is a true negative result rather than a failure of the process. Positive controls can be diluted plasmids containing the cDNA of interest, but these need to be kept separate from experimental samples, preferably in a completely different laboratory, to ensure that cross-contamination does not occur.

20.3 Common problems and pitfalls

20.3.1 Primer design

Although nucleic sequences of 15–30 bases occur relatively infrequently in the transcriptome, there is still potential for mis-hybridization (false annealing) if primer sequences are not specific to the sequence to be identified. All primer sequences should be checked against the sequence databases for this problem (most people use an algorithm called BLAST: see www.ncbi.nlm.nih.gov/BLAST/).

Primer design dictates the temperature at which primers anneal to target DNA. As a general rule, to give appropriate specificity, primers should have similar annealing temperatures ($\pm 2°C$), equal GC content (between 40–60 per cent), should not be self-complementary (as they form secondary structure), especially in the 3' region, and should be longer rather than shorter.

There are many treatise written on primer design. Primers can be designed to introduce alternative sequences, often to introduce restriction enzyme recognition sites to enable identification or cloning. There are many free online tools available to aid primer design, and one popular one is Primer3 at http://frodo.wi.mit.edu/. However, there are many others that also do good jobs.

20.3.2 Verification of amplicon identity

DNA from a PCR reaction that runs on a gel at approximately the correct (expected) length is *not* sufficient evidence that one has amplified the correct target. Agarose gel electrophoresis cannot accurately resolve DNA bands of lengths that differ by 50 nucleotides, so amplicon length is only a rough guide to identity. Amplicon identity should be confirmed by direct sequence analysis, restriction mapping, or by Southern blotting of the gel.

20.3.3 Lack of PCR product

This is probably the most common problem with PCR, and the one that is often most difficult to troubleshoot. Inclusion of appropriate positive and negative controls will tell you whether this is a true negative or an experimental failure. Lack of amplification could indicate a number of different problems, ranging from poor template, poor primers, nuclease contamination, inadequate ionic concentrations (e.g. Mg^{2+} concentration), or the presence of inhibitors of PCR. For a comprehensive troubleshooting guide, see Altshuler (2006).

20.3.4 Reaction conditions

Experience helps, of course, but things to watch out for include whether the constituents of the reaction are limiting in concentration (nM) or in excess (mM) – they need to generally be in excess although too much in excess (e.g. template) can result in lack of amplification. Some things to check are:

- Are incubation times short enough to ensure specificity of annealing ($\approx$30 seconds) but long enough to ensure adequate extension of the DNA molecule (1 minute per kilobase of target)?

- Is the cycle number sufficient, but not too many (high cycle numbers tend towards reduced specificity)?

20.3.5 Presentation of results

The whole gel should be shown for a PCR reaction to enable the reviewer to see whether extraneous amplicons have been generated. In general, only one band should be seen on a gel. Explanations should be offered for multiple bands (e.g. previously unknown splice variants of the same gene should be identified to ensure that they are 'real' and not a result of mispriming and amplification of unrelated products).

20.3.6 Subsequent use of the DNA

If the DNA produced by PCR is then used in expression studies, or as a probe for examination of, for example, mRNA levels, it is absolutely vital that the identity of the PCR product is confirmed. It is also important that the correct DNA polymerase is used; for cloning and further expression, it is best to use a proofreading enzyme (e.g. Pfu) rather than standard Taq polymerase, because enzymes such as Pfu exhibit greater fidelity of polymerization (approximately 10 fold lower error rate, for example 8 ± 3.9 compared to 1.3 ± 0.2 (error rates $\times 10^{-6}$, $\pm$ SD) (Cline *et al.*, 1996).

20.3.7   Quantification of DNA product

One of the biggest problems in the use of RT-PCR and real time-PCR is the quantification of product. PCR is an amplification reaction, and not all PCR products are amplified with the same efficiency. Smaller products will be amplified more rapidly, and will exhaust reaction components such as primers more rapidly than larger products. Within the amplification phase, the amplification can be described by the following equation (Becker-Andre, 1995):

$$\text{Amplification} = \text{amount input cDNA}.(1 + \text{efficiency})^{\text{number of cycles}}$$

In order to relate different reactions to each other, the amplification has to be the same and, in order to do this, the amount of input target *and* the efficiency of the reaction have to be exactly the same. In reality, the reproducibility of the efficiency of the reactions between tubes is nowhere near high enough to be able to quantitate the product accurately. There are, however, ways in which this can be overcome. Standards can be included in the RNA sample (competitive PCR methods such as PATTY – polymerase chain reaction-aided transcript titration assay) (Becker-Andre, 1995), limited dilution PCR (Sykes & Morley, 1995) or by looking at relative, rather than absolute transcript expression (Larrick & Siebert, 1995).

In many 'quantitative' methods, comparison is made to expression of mRNA encoding so-called housekeeping genes such as GAPDH or β-actin, but it should be remembered that these mRNAs may also alter expression level. It is therefore wise to include comparison to more than one such control, in case expression of the 'control' also changes.

Real time-PCR will not be dealt with in any detail here. Quantitation of real time-PCR product can be done and, indeed, this technique is now very often used for exactly this purpose. However, the best methods, and the accuracies of quantification, are the subject of much discussion. A mathematical analysis and discussion of quantitation of real time-PCR can be found in Page & Stromberg (2011).

20.4   Complementary techniques

Primer 13: Immunocytochemistry.
Primer 14: Immunoprecipitation.
Primer 15: Immunoblotting.
Primer 23: Microarray (gene chip).

Further reading and references

Altshuler, M.L. (2006). *PCR Troubleshooting: The Essential Guide*. Caister Academic Press, Wymondham, Norfolk, UK.

Becker-Andre, M. (1995). In: Larrick J.W. & Siebert P.D. (Eds.) *Reverse transcription PCR*, pp. 121–149. Ellis Horwood Ltd, UK.

Cline, J., Braman, J.C. & Hogrefe, H.H. (1996). PCR fidelity of pfu DNA polymerase and other thermostable DNA polymerases. *Nucleic acids research* **24**, 3546–3551.

Donaldson LF, Humphrey PS, Oldfield S, Giblett S, Grubb BD. (2001). Expression and regulation of prostaglandin E receptor subtype mRNAs in rat sensory ganglia and spinal cord in response to peripheral inflammation. *Prostaglandins and Other Lipid Mediators* **63**(3), 109–122.

Larrick J.W. & Siebert P.D. (Eds.), (1995) *Reverse transcription PCR*. Ellis Horwood Ltd, UK.

Page, R.B. & Stromberg, A.J. (2011). Linear methods for analysis and quality control of relative expression ratios from quantitative real-time polymerase chain reaction experiments. *The Scientific World Journal* **11**, 1383–1393.

Sykes, P.J. & Morley, A.A. (1995). In: Larrick J.W. & Siebert P.D. (Eds.) *Reverse transcription PCR*, pp. 150–165. Ellis Horwood Ltd, UK.

Taylor, G.R. (1991). Polymerase chain reaction: basic principles and automation. In: McPherson, M.J., Quirke, P. & Taylor, G.R. (Eds.) *PCR – A practical approach*, vol. 1, pp. 1–14, Oxford University Press, Oxford.

21
In Situ Hybridisation (ISH)

Lucy F. Donaldson
Physiology & Pharmacology, University of Bristol, UK

21.1 Basic 'how-to-do' and 'why-do' section

In situ hybridization (ISH) is used to detect nucleic acid in fixed tissue, using a labelled nucleic acid probe to bind to the target through Watson-Crick base-pairing. The target can be chromosomal DNA, RNA or infective viral DNA/RNA. This primer will concentrate on the localization of RNA, which is the most commonly used permutation of ISH in physiology and neuroscience. ISH allows one to examine the spatial and temporal resolution of nucleic acid localization; this is particularly applicable to the study of RNA, as one can gain insight into the developmental expression of specific gene products, as well as the cell/tissue types in which it is found. In addition, ISH can be used to quantify the amount of nucleic acid when semi-quantitative techniques are used. Thus, ISH is used to determine when, where and how much RNA is being expressed.

ISH is a multi-step technique that requires significant optimization; there are many different published protocols in the literature. Here I will describe a generic approach, but bear in mind that there may be differences between this method and those published elsewhere that may not be entirely obvious. There are detailed protocols available in the literature (for detailed methods to localize low abundance RNAs, see Donaldson (2004)). ISH can be combined with other techniques, for example immuno-histochemistry for co-localization of protein and mRNA (Newton *et al.*, 2002).

Step 1 Probes are labelled, often by incorporation of a radioisotope such as ^{35}S. Radiolabel is usually necessary if a semi-quantitative or quantitative approach is needed. Labelling can be achieved either by incorporation of isotopically labelled nucleotides into the body of the molecule (using PCR or *in vitro* transcription) or by 'tailing' a molecule with several isotopically labelled nucleotides (end-labelling).

Essential Guide to Reading Biomedical Papers: Recognising and Interpreting Best Practice, First Edition.
Edited by Phil Langton.
© 2013 by John Wiley & Sons, Ltd. Published 2013 by John Wiley & Sons, Ltd.

Non-radioactive labels can also be used, and detection of these is similar to detection of signal in immunocytochemistry and immunofluorescence.

Step 2 Tissue sections mounted on slides are first fixed with a cross-linking fixative (this treatment can effectively inactivate any RNA degrading enzymes, RNases (Tongiorgi *et al.*, 1998)), washed to remove fixative and hybridized to the probe. Tissue is often dissected without fixation, snap-frozen to prevent the action of tissue RNAses and the degradation of RNA, and stored at $-80°C$. Frozen tissue sections are usually thin ($<10\ \mu m$) to enable accurate localization, and can be stored mounted on slides at $-80°C$ until required.

The probe is added to the tissue sections in excess, to ensure that all target mRNA is hybridized to the probe. This is very important if RNA is to be quantified. Hybridization is usually done overnight at fairly high temperatures (usually 50–60°C) in a humidified atmosphere. Hybridization temperature should be optimized – a recommended starting point is probe melting temperature (Tm) minus 20°C. For further details on this, see Donaldson (2004). High temperatures are used for hybridization to minimize non-specific binding (hybridization is the formation of macromolecules by recombination of complementary subunits (i.e. the probe and the target)).

Step 3 Non-specifically hybridized (weakly bound) probe is washed off. Nucleic acid probes can stick to non-complementary targets, or just to 'sticky' areas within the tissue. Binding of probes to non-complementary targets is destabilized and therefore removed by washing under 'stringent' conditions. High stringency conditions are those under which only fully complementary targets are stable, and all other probe/tissue interactions are unstable and therefore removed by the washing. Stringency is increased by increasing temperature (hence the reasonably high hybridization temperature), decreasing sodium concentration and increasing concentrations of chemicals, such as formamide, that destabilize hydrogen bonds.

Step 4 Radioactive signal is localized. This is achieved at the whole-tissue level by putting tissue sections directly against x-ray film. The radioactive emission from the areas containing probe and target (hybrids) causes a blackening in the developed film. This method gives an overview of tissue distribution, but it cannot give the resolution required for precise information about cellular distribution of hybridization.

For cellular information, the tissue sections are coated with a liquid photographic emulsion applied directly onto the surface of the tissue. Radioactive emission then precipitates silver grains in the layer of photographic emulsion directly over any cells containing stable probe/target hybrids. Development of the emulsion allows visualization of the precipitated silver grains, or leaves a transparent emulsion layer over areas of tissue with no radioisotope present. Either of these methods gives information on the tissue distribution of the RNA target under study.

Step 5 Quantitation of the signal; the amount of RNA relative to a control condition or tissue can be determined using *in situ* hybridization with either of the above methods. With film autoradiography, the optical density of the film image can be calculated and compared between control and experimental situations. With emulsion autoradiography, because the amount of probe/target is directly related to the number of precipitated silver particles, these can be counted to give a value for the relative RNA level in an area of tissue, or more powerfully at the single cell level (i.e. this cell has more/less mRNA than that cell). Both of these methods are semi-quantitative, i.e. a comparison between a control and an experimental situation. Absolute quantitation can be performed by either method by incorporation of standards with known isotopic specific activity to generate optical density/silver particle number – isotope/RNA amount relationships.

There are many different protocols available for non-isotopic ISH. Many different molecules, such as biotin (as used for immunohistochemistry) or digoxygenin have been incorporated into probes for ISH. Fluorescent labels can also be used, but have been primarily applied in the area of FISH (Fluorescent *In Situ* Hybridization) that usually refers to the localization of DNA to specific areas of chromosomes.

21.2 Required controls

Controls for ISH are hotly debated. There is no one 'gold standard' control that can be used; many experiments cannot be interpreted with confidence. It is usual to find at least two of the following controls (all have their limitations).

1. Hybridization of the tissue with a 'sense' probe, i.e. a probe of the same (as opposed to the complementary) sequence to the target. This is a commonly used and requested control. **Aim**: to control for non-specific probe binding to the tissue due to something in that specific sequence. **Major problem**: It is now known that many genes are transcribed from *both* DNA strands, and therefore a sense probe can generate a signal by specific binding to a related and functional RNA.

2. Pre-digestion of target RNA. **Aim**: to demonstrate that signal is associated with binding of the probe to RNA in the tissue (if there is no target, then there should be no signal). **Problem**: shows the signal is derived from binding of the probe to RNA(s), but not necessarily that the binding is specific to the RNA of interest, because RNase treatment of the tissue will degrade all RNA in the tissue, not just the target.

3. Use of multiple probes aimed at different parts of the same target. **Aim**: should show that all probes have the same distribution of signal. **Problem**: now that it is recognized that many genes give rise to multiple splice variants, probes from different areas of the predicted mRNA sequence that give different distributions may be detecting slice variant differences.

4. Probes targeted to different members of the same gene family or to other related molecules. **Aim**: should give different distributions of signal in the same tissue. **Problem**: this control does not, of course, verify specificity of binding of the target probe.

5. Competition for hybridization with excess unlabelled sense RNA. **Aim**: should give no signal; excess unlabelled RNA will competitively inhibit the binding of labelled probe. **Problem**: this control tells you that your probe binds to that sequence in solution and in isolation. It does not tell you whether it *also* binds to an alternative sequence which might be present in your tissue.

6. Increasing stringency washes. **Aim**: washing the slides at very high stringency should completely remove the probe, resulting in no signal. **Problem**: if the signal is very strong after high stringency washes, it is likely that non-specific interactions are present.

7. Hybridization of the probe against tissue from knockout animals. **Problem**: the biggest problem with this control is if the method used to disable the gene of interest continues to allow mRNA transcription, for example of a partial (truncated) transcript due to exon deletion, which is not translated, then ISH signal may still be detected, even though protein expression will not be present.

21.3   Common problems and pitfalls in execution or interpretation

21.3.1   Is the signal specific?

Non-specific signal can arise for many reasons. Things to be aware of are:

- *How long are the probes used?* Short probes (<100 nucleotides) can be less specific than long probes (>150 nucleotides), due to the probability of short sequences occurring frequently in related genes.

- *Are the probes DNA or RNA?* RNA/RNA hybrids are more stable than RNA/DNA hybrids and therefore need higher stringency conditions to be sure of specificity.

- *Is stringency of hybridization and washing high enough?* With any probe, if hybridization and washing are done at too low a temperature, some signal may be non-specific. With RNA probes, hybridization temperatures should be 40–60°C in the presence of ≈50 per cent formamide and other destabilizing agents. Hybridization temperatures lower than this, lower formamide concentrations, or higher sodium concentrations during hybridization, can lead to high background. While it is tempting to use very low, non-stringent hybridization conditions just to get a signal, it can sometimes be very difficult to remove non-specific binding under these conditions.

21.3.2   Background problems

Some background labelling can be attributable to the problems mentioned above. Background labelling can also occur if probes are designed against common sequences (areas that are very GC rich, for example). With ^{35}S-labelled probes, oxidation can result in non-specific binding, so all steps should be done in the presence of a reducing agent (e.g. β-mercaptoethanol or dithiothreitol). In radio-active *in situ* hybridization, background also increases with longer exposure times without increasing signal intensity.

21.3.3   Protein versus mRNA

A common source of debate around ISH occurs when there is a cellular mismatch in protein and mRNA localization. Some points to consider are:

- mRNA can be found in the absence of protein (this is *very rare*, but often the argument is used in defence of such findings). In contrast, a cell expressing a particular protein *must* have expressed the mRNA at some point, or the protein could not have been translated.

- mRNA and protein can have very different time courses of expression (generally, mRNA is temporally ahead of protein) and, as a result, the time point of the study may find one and not the other. mRNA and protein can also have very different half-lives in different tissues; some mRNAs are rapidly degraded, while some proteins turn over very slowly.

- Nucleic acid hybridization techniques at high stringency are extremely specific, especially when using longer probes. This is not always true of antibodies; although antibody/antigen interactions are highly specific, in raising an antibody it is often not possible to be sure that it is not cross-reacting with a completely different protein. The probability of this happening in a nucleic acid hybridization reaction can be calculated, and is 1 chance in 4 to the power probe length (1 in $4^{150} =$ very, very, very unlikely!).

21.3.4   Contamination and true negative results

As with any experiment, a negative result can always be attributable to an experiment just not working, rather than what you are looking for not being there. An important and recurring message in this book is that *absence of evidence does not equate to evidence of absence*. Negative results should always be verified by inclusion of a positive control, especially when working with RNA, which is very easily and rapidly degraded.

21.4 Complementary and/or adjunct techniques

Primer 13: Immunohistochemistry.
Primer 14: Immunoprecipitation.
Primer 15: Immunoblotting.
Primer 23: Microarray (gene chip).

Further reading and resources

Donaldson, L.F. (2004). Identification of G-protein-coupled receptor mRNA expression by Northern blotting and in situ hybridization. *Methods in Molecular Biology* **259**, 99–122.

Newton, S.S., Dow, A., Terwilliger, R. & Duman R. (2002). A simplified method for combined immunohistochemistry and in-situ hybridization in fresh-frozen, cryocut mouse brain sections. *Brain Research. Brain Research Protocols* **9**, 214–219.

Tongiorgi, E., Righi, M. & Cattaneo, A. (1998). A non-radioactive in situ hybridization method that does not require RNAse-free conditions. *Journal of Neuroscience Methods* **85**, 129–139.

Wilkinson, D.G. (1998). *In situ hybridization: a practical approach*, 2nd ed. Practical Approach Series (Oxford University Press).

22
Methods of Interference (Antisense, siRNAs and Dominant Negative Mutations)

Allison Fulford

Centre for Comparative and Clinical Anatomy, University of Bristol, UK

22.1 Basic 'how-to-do' and 'why-do' section

22.1.1 General uses of interference techniques

In order to study the function of genes in a population of cells or tissue, one can adopt a traditional pharmacological approach or a molecular approach. A pharmacological strategy depends on the availability of selective drugs or ligands to influence directly actions of cellular enzymes or receptor proteins *in vitro* or *in vivo*. An alternative and increasingly popular molecular approach involves manipulation of cellular RNA molecules. Synthetic nucleic acid molecules are used to disrupt endogenous expression of target genes in cells and tissues. This strategy aims to eliminate (in practice, reduce) production of native, endogenous peptides or proteins either by blocking specific mRNAs using antisense mRNA or by gene silencing using RNA interference (RNAi). These 'interference' techniques can be used to characterize the function of unknown genes using loss-of-function analysis.

Collectively, these methods employ oligonucleotides to 'knockdown' gene function *in vitro* or *in vivo*. This general approach has only been an option in the 'post-genomic', era as coding information for the entire genome of several important experimental species has become publically available. Interference

Essential Guide to Reading Biomedical Papers: Recognising and Interpreting Best Practice, First Edition.
Edited by Phil Langton.
© 2013 by John Wiley & Sons, Ltd. Published 2013 by John Wiley & Sons, Ltd.

oligonucleotides are powerful tools in the biologist's armoury, and advances in research may yet prove that some have therapeutic potential.

22.1.2   Implications of this approach

- Novel genes with unknown roles can be functionally characterized by manipulation of levels of endogenous RNAs.

- Antisense interference can be used to eliminate production of all forms of endogenous peptide which may be translated from a particular mRNA. This is relevant when differential processing of a precursor protein between tissues may produce tissue-specific isoforms of products. However, the effectiveness of antisense oligos is dependent on the relative stability and specific half-lives of target protein/peptides.

- Highly selective and safe drugs based on antisense or RNAi technology may eventually be developed, with improved specificity and lower toxicity than conventional drugs (e.g. for cancer, immunological or genetic conditions).

22.2   Types of interference

Interference techniques have enormous scope, given the multitude of cellular proteins available for functional characterization. The general characteristics of the main types of interference approach are outlined below.

22.2.1   Antisense

The theory behind the use of antisense oligonucleotides is that production of specific proteins can be prevented by inhibition of translation of particular mRNAs. A number of different approaches have been attempted to inhibit mRNA translation using antisense technology.

Antisense oligonucleotides (oligos) are short, single-stranded sequences of RNA or DNA that are designed to bind to, or hybridize with, their complementary mRNA molecules in the host. The hybrid formation causes a steric or conformational change that blocks translation of the mRNA, ultimately preventing synthesis of protein in the cytoplasm of the cell. In an ideal result, production of a specific protein is inhibited without affecting the translation of other mRNAs. An encoded protein product's function can therefore be inferred from the loss-of-function phenotype.

Antisense oligos are usually chemically modified to enhance stability. For example, phosphorothioate oligos have a sulphated DNA backbone that reduces degradation. Alternatively, unprotected phosphodiester oligodeoxynucleotide antisense probes have been demonstrated to have specificity and efficacy *in vitro* without the unwanted cellular toxicity associated with some modified oligonucleotides.

22.2.2 RNA interference

RNA interference is a more recently developed technique that exploits the cell's own ability to manufacture RNA-interfering molecules that enable endogenous gene silencing. RNAi technique can be used to silence the expression of target genes in a variety of organisms and cell types (from both animals and plants), both *in vitro* and *in vivo*. RNAi is a technique in which exogenous double-stranded RNAs (dsRNA) that are complementary to known messenger RNAs (mRNA) of the species of interest are introduced into a cell. These dsRNAs stimulate a cellular response, resulting in cleavage of the target mRNA and thereby decreasing or abolishing protein synthesis. Originally characterized in the roundworm, *Caenorhabditis elegans*, RNAi has also been demonstrated in mammalian cells in culture. Scientists have attempted to induce RNAi in mammals *in vivo* with some success, but much more research is required in this area, given the difficulties in intracellular delivery of oligonucleotides in living animals.

In recent years, a variety of endogenous types of small RNA species have been discovered that are vital for post-transcriptional processing of RNAs in cells (see Singh, 2011). These include small interfering RNAs (a type of dsRNA), small hairpin RNAs and microRNAs. MicroRNAs are non-coding RNAs that are ubiquitous in nature. They are important for targeted control of protein synthesis, owing to their regulatory role in gene transcription. They represent another mechanism for inhibition of gene expression, and abnormalities in their regulation are implicated in cancer development.

Methodology for design and delivery of interference oligonucleotides The experimental approach for using antisense and RNAi is similar. The most formidable challenge is enabling cell uptake of the synthetic oligonucleotides and, subsequently, confirming appropriate targeting to the gene of interest. In all cases, molecules should be fully characterized *in vitro* before attempting to undertake *in vivo* experiments, if the latter is the ultimate objective.

The targeted delivery of oligonucleotides is a challenging aspect of interference techniques. For successful manipulation of gene expression using antisense or RNAi, synthetic oligonucleotides must be taken up into cells. In RNAi, small interfering RNAs (siRNA) are utilized. SiRNAs act as intermediates in the RNA interference pathway, ultimately leading to specific silencing of genes in somatic cells. They are short dsRNA molecules, usually between 21–25 nucleotides in length, that can be chemically synthesized or prepared by *in vitro* transcription.

Use of siRNAs in mammalian systems is essential, as longer dsRNAs trigger a non-specific Type 1 interferon response leading to cell death, whereas siRNAs appear to bypass this immune activation and induce sequence-specific gene silencing. This goal depends on the uptake of the small RNAs into the cell and incorporation into the cell's own **RNA induced silencing complex** (RISC). This nuclease complex uses the siRNA sequence as a guide to target homologous RNAs. Once RISC binds to its target mRNA strand, the latter is cleaved.

Nucleic acids, being large polar (hydrophilic) molecules, do not readily diffuse across cell membranes, so methods of delivery need to be employed. Delivery methods need to be chosen carefully and may involve the use of cationic lipids, polymers (e.g. polyethylenimine), electroporation or siRNA expression vectors. Methods for delivery are highly related to the cell type under investigation, and some types of cells are easier to manipulate than others.

Neuroscientists are particularly interested in harnessing RNAi technology for studies of mammalian neurones. However, difficulties exist, as neurones are notoriously difficult to transfect *in vitro* and *in vivo*, and thus viral vectors (see Primer 19) may be required to ensure a satisfactory delivery of oligonucleotide to the intracellular compartment of the neurone (e.g. herpes simplex viral vector, adeno-associated viral vector).

Experimental designs for interference molecules

A. **Basic *in vitro* assay**

1. Establish cell line or primary cell culture pertinent to the study.

2. Deliver antisense or siRNA to cells in culture over a dose range (or add appropriate negative control – see below), either alone or in the presence of a transfection reagent to enhance uptake.

3. Culture cells for a defined time period (e.g. 12, 18, 24, 48 and 72 hours) to determine time-course of effect.

4. Determine cellular uptake efficiency of antisense/siRNA treatment.

5. Determine effects of treatment on physiological end point, e.g. cell proliferation, growth, death or other parameter.

6. Analyze treatment effects versus control using appropriate statistical analysis.

B. ***In vivo* study**

1. Identify appropriate test species or animal disease model.

2. Determine delivery method, e.g. systemic or CNS injection or co-administration using transfection reagent.

3. Type of experiment? Acute (single injection), chronic (repeated injection) or continuous infusion.

4. Administer antisense or siRNA (or appropriate negative control) to animal over selected dose range.

5. Measure physiological end-point or monitor behaviour over defined time-period (e.g. 12, 24, 48, 72 hours).

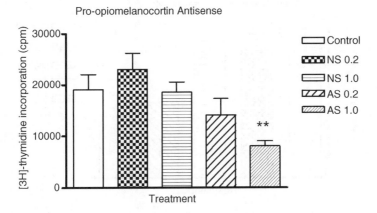

Figure 22.1 Effect of an antisense oligodeoxynucleotide complementary to pro-opiomelano-cortin (POMC) mRNA on rat immunocyte proliferation *in vitro*. ^{3}H-thymidine uptake by rat spleen immunocytes (splenocytes) cultured with 0.2 or 1.0 µmol/l antisense (AS) or non-sense (NS) pro-opiomelanocortin (POMC) phosphodiester oligo probes for 24 hours prior to stimulation with mitogen, Concanavalin A (5 µg/ml) for 72 hours. Values are mean ± SEM, $n = 6$. **$p < 0.01$ compared with controls (no probe), one-way ANOVA with posthoc Fisher PLSD test. ^{3}H-thymidine uptake provides a index of cell proliferation. Incubation of rat splenocytes with a POMC antisense probe (1.0 µmol/l) for 72 hours results in a reliable and reproducible inhibition of cell activation, as measured in terms of mitogen-stimulated cell proliferation. Oligos were 18 base-length antisense or non-sense probes and designed 'in-house'. Courtesy of Dr. Alison Fulford.

6. Terminate experiment and determine cellular uptake efficiency of treatment.

7. Analyze treatment effects versus control using appropriate statistical analysis.

In the example of a basic type of antisense experiment (see Figures 22.1 and 22.2), antisense effects on cell proliferation have been studied in a primary cell culture. This approach has compared antisense versus non-sense probes at two different concentrations and, in addition, has validated the antisense by quantification of antisense-induced changes in production of the POMC-encoded product, β-endorphin. Limitations of the approach are the incomplete inhibition achieved by the antisense oligo and the absence of cellular uptake information to confirm appropriate distribution of the oligo to the intracellular compartment of the splenocyte.

22.2.3 Dominant negative mutations (DNMs)

Dominant negative inhibition is a powerful genetic approach for characterization of gene function *in vivo*. DNMs rarely occur in nature, but they can be engineered to inhibit the function of newly expressed native protein via over-expression of the mutant version in cells or tissues (see Sheppard, 1994). DNMs can cause functional inhibition via several mechanisms, including production of a defective protein, production of a target protein that blocks function of a native protein, or production of protein that stably binds to a DNA regulatory motif.

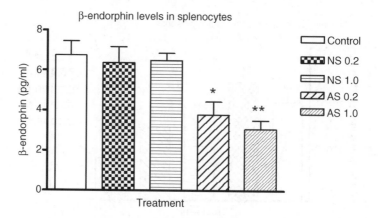

Figure 22.2   Effect of an antisense oligodeoxynucleotide complementary to pro-opiomelano-cortin (POMC) mRNA on β-endorphin immunoreactivity *in vitro*. β-endorphin immunoreactivity measured by radioimmunoassay in splenocytes cultured for 24h with 0.2 or 1.0 μmol/l antisense (AS) or non-sense (NS) pro-opiomelanocortin (POMC) phosphodiester oligo probes. Cells were stimulated with Concanavalin A (5 μg/ml) for 72 hours. Values are means ± SEM of ir-β-endorphin assayed in triplicate. β-endorphin is a POMC-derived peptide synthesized in rat spleen. Incubation of splenocytes for 24 hours with the POMC antisense probe decreased the amount of ir-β-endorphin in the cells by more than 55 per cent compared to cells incubated with non-sense or control. *P < 0.05, **P < 0.01 versus controls and nonsense probe, one-way ANOVA with posthoc Fisher PLSD test. Courtesy of Dr. Alison Fulford.

DNMs can be used to manipulate cellular substrates acted upon by native proteins, whereby a mutant protein can be over-expressed that binds more effectively to a key substrate, thus competitively inhibiting function of the native protein. DNMs are also an effective way of impeding function of multi-subunit proteins, such as receptor dimers, which depend on the formation of an active heterodimer complex involving independent protein subunits.

Expression of a DNM in a tissue enables non-functioning dimeric receptors to be expressed that nullify the normal cell response to an endogenous ligand. Inhibition can occur by several mechanisms, including formation of an inactive heterodimer complex. DNM have been successful in the analysis of multimeric proteins, oncogene products, hormone receptors and growth-factor receptors. The expression of DNMs under the control of tissue-specific promoters holds promise for the study of the function of novel proteins.

Methodology for design and delivery of DNM constructs   The experimental approach is complex and involves engineering of the cDNA of the target protein. This involves the generation of a DNM transgene construct encoding a mutant protein (e.g. a truncated variant), with a promoter that will effectively drive expression. The transgene is introduced into an appropriate model *in vitro* or *in vivo*, using targeting vector delivery, e.g. adenoviral vector (Flynn *et al.*, 2004). Control transgene constructs containing the full-length coding region of the cDNA and the same promoter are used as negative controls. Gene function can be inferred from the analysis of the phenotype of the genetically-modified organism. Confirmation of change in the expressed level of the mutant protein, compared to wild-type protein, needs to be assessed using immunoblotting (see Primer 15).

22.3   Required controls and problems or errors in literature

There are a number of potential problems encountered when using interference techniques.

22.3.1   Antisense and RNA interference

Positive and negative controls should be routinely employed. Positive controls are based on using more than one antisense oligo or siRNA. Additional sequences that target the same mRNA should induce the same effect, thus confirming specificity of the reducing effect. Negative controls are also very important. A negative control sequence contains the random base sequence (non-sense) or scrambled nucleotide composition of the test antisense or siRNA, but lacks any homology to the genome. Homology can be checked by comparison of query sequences with known DNA sequence information logged in databases such as the NIH genetic sequence database (GenBank).

Validation experiments must confirm that the antisense oligo or siRNA molecule enters the appropriate part of the cell. Fluorescent labelling of antisense or siRNA molecules is an additional control to monitor their correct cellular uptake and distribution. A problem in the literature is that often the validatory work, such as extensive checks that the oligo cellular uptake is appropriate, may not be reported – or worse, it may not even be attempted. Other concerns may be that DNA sequence information of oligos is not sufficiently detailed in the reports. This information is essential and may be appended to online supplementary information.

Any technique based on inhibition of mRNA translation should confirm that mRNA levels have been reduced (by Northern blotting or RT-PCR – see Primer 20). Observations of biological effects attributed to antisense or siRNAs should be confirmed by the demonstration of changes in target protein level (by immunoblotting – see Primer 15).

22.3.2   Dominant negative mutations

Use of control constructs encoding the native protein must be used. Confirmation of change in the expressed level of the mutant protein, compared to wild-type protein, must be shown by immunoblotting. Furthermore, suitable reagents (e.g. probes, antibodies) should be available to detect the expression of the transgene *in vivo*.

22.4 Common problems and pitfalls in execution or interpretation

22.4.1 Antisense and RNA Interference

Design Accessible sites of target mRNA for oligo binding or siRNA activity must be identified, and the initial optimization experiments required are both expensive and time-consuming. More efficient cellular uptake is achieved with shorter oligos (e.g. 18–25 base-pairs long).

Specificity of effect In some cases, biological effects of oligos (e.g. on phenotype or protein level) may be seen using scrambled sequence controls, suggesting a degree of off-target effect of the oligo/siRNA. Alternatives should be sought that lack these non-specific effects on gene expression. All sequences must be cross-checked against gene sequence databases to ensure no appreciable homology with known genes. Confirmation of active antisense/siRNA-induced changes in gene expression and protein level must be confirmed by northern- and immuno-blotting respectively.

Duration and magnitude of response The duration and magnitude of response is related to the dose, stability and pharmacokinetics of the oligo or siRNA. Dose-response experiments are vital to determine the minimum concentration of antisense/siRNA required to induce maximum inhibition, whereas time-course studies are necessary to calculate the half-life of the oligo/siRNA. For antisense experiments, the degree of inhibition caused by most antisense oligos ranges from 50–80 per cent whereas, with RNAi, the degree of inhibition is higher (75–95 per cent). For this reason, RNAi has become the favoured method for gene silencing. The maximum silencing effect is dependent on a number of factors, including the expression level of the target mRNA, the half-life of the target protein and the uptake efficiency of the antisense/siRNA.

Delivery problems Methods for delivery of oligos/siRNAs will be dependent on whether the experiment is an *in vitro* or *in vivo* study. Development of systems for efficient delivery of RNAs is highly time-consuming and can be difficult to optimize. There are a number of non-viral and viral methods for targeted nucleic acid delivery.

For *in vitro* assays, different populations of cells respond differently to transfection methods (e.g. some are more susceptible to viral vectors than others). In addition, adherent cells and suspension cell cultures differ in their sensitivity to transfection reagents, with suspension cultures requiring electroporation-based delivery.

Delivery of oligos/siRNAs *in vivo* is especially problematic, with challenges in terms of overcoming issues with cellular uptake, organ delivery, metabolic stability and cellular toxicity. Some success has been reported using transfection reagents, lipid-based carriers and viral vector delivery systems, especially with regard to delivery of phosphorothioate antisense oligonucleotides. SiRNA-based drugs are especially challenging for *in vivo* use, as they do not tolerate structural modifications well, and this can prevent interaction with the RISC. They are also highly unstable when given systemically. However, clinical trials have demonstrated that they can be delivered locally via the intranasal route and they are successfully taken up by the liver when delivered by lipid-based carriers.

Toxicity   Both oligos and siRNAs can be associated with toxic side-effects or immunogenic responses, especially at high concentrations. Certain chemical modifications of antisense oligos have been associated with higher cellular toxicity. For both antisense and RNAi, it is important that the minimal concentration that produces a biological effect is used in experiments in order to minimize the possibility of side-effects. Chemically-synthesized siRNAs appear to be less toxic than *in vitro*-transcribed siRNAs.

22.4.2   Dominant negative mutations

Some experiments have suggested that dominant-negative proteins are more rapidly degraded than wild-type proteins. This may explain the absence of any significant change in phenotype *in vivo* seen in some studies.

Testing DNM in animals is a costly and labour-intensive process. DNMs should be induced in *in vitro* assays first to confirm transgene functionality.

22.5   Complementary and/or adjunct techniques

Primer 14: Immunoprecipitation.
Primer 15: Immunoblotting.
Primer 20: PCR.
Primer 21: *In situ* hybridization.

Further reading and resources

Flynn, A., Whittington, H., Goffin, V., Uney, J. & Norman M. (2004). A mutant receptor with enhanced dominant-negative activity for the blockade of human prolactin signalling. *Journal of Molecular Endocrinology* **32**, 385–396.

Galderisi, U., Cascino, A. & Giordano, A. (1999). Antisense oligonucleotides as therapeutic agents. *Journal of Cellular Physiology* **181**, 251–257.

Herskowitz, I. (1987). Functional inactivation of genes by dominant negative mutations. *Nature*, **329**, 219–222.

Jones, S.W., Souza, P.M. & Lindsay, M.A. (2004). SiRNA for gene silencing: a route to drug target discovery. *Current Opinion in Pharmacology* **4**, 522–527.

Kanasty, R.L., Whitehead, K.A., Vegas, A.J. & Anderson, D.G. (2012). Action and Reaction: The biological response to siRNA and its delivery vehicles. *Molecular Therapy* **20**, 513–524.

Kole, R., Krainer, A.R. & Altman, S. (2012). RNA therapeutics: beyond RNA interference and antisense oligonucleotides. *Nature Reviews Drug Discovery* **11**, 125–140.

Phillips, M.I. (2000). Antisense technology Part A, General Methods, Methods of Delivery & RNA studies. *Methods in Enzymology* **313**, Academic Press.

Sheppard, D. (1994). Dominant Negative Mutants: Tools for the Study of Protein Function *In Vitro* and *In Vivo*. *American Journal of Respiratory Cell and Molecular Biology* **11**, 1–6.

Singh, S., Narang, A.S. & Mahato, R.I. (2011). Subcellular fate and off-target effects of siRNA, shRNA, and miRNA. *Pharmaceutical Research* **28**(12), 2996–3015.

Zalachoras, I., Evers, M.M., van Roon-Mom, W.M., Aartsma-Rus, A.M. & Meijer, O.C. (2011). Antisense-mediated RNA targeting: versatile and expedient genetic manipulation in the brain. *Frontiers in Molecular Neuroscience* **4**: 10.

23
Transcriptome Analysis: Microarrays

Charles Hindmarch

Clinical Sciences, University of Bristol, UK

23.1 Basic 'how-to-do' and 'why-do' section

Investigations into the genomic response to environmental fluctuation or pathology have previously been limited to the weight of supporting literature, researcher intuition and the availability of appropriate probes. Microarray allows a non-biased approach to identifying which genes change their expression in a particular tissue or cell type in response to a given physiological challenge.

A microarray is a technology that allows simultaneous measurement of the expression of hundreds, thousands or tens of thousands of genes. In its simplest form, a microarray is a library of cDNA or oligonucleotide probes that have been immobilized onto a substrate such as nylon, glass or quartz. Each of these probes is a sequence that will hybridize to a specific and known messenger ribonucleic acid (mRNA) sequence according to Crick-Watson base pair rules – one probe, one gene.

Because microarray interrogates the [1]*transcriptome*, the first step is to extract and purify RNA from each sample. Total RNA is composed of different populations, of which only ≈5 per cent is considered to be coding (mRNA), with the remaining fractions being ribosomal and non-coding small RNA species such as micro-RNAs. Given that the total RNA concentration of a single cell is in the low picogram range, obtaining sufficient mRNA to perform transcriptomic analysis can be a challenge. However, amplification protocols address these needs. Selective amplification of mRNA during reverse transcription is ensured through the use of specialized

---

[1] **Transcriptome** – all mRNA transcripts from the sample.

Essential Guide to Reading Biomedical Papers: Recognising and Interpreting Best Practice, First Edition. Edited by Phil Langton.
© 2013 by John Wiley & Sons, Ltd. Published 2013 by John Wiley & Sons, Ltd.

primers bearing a T7 promoter region that binds to the polyadenylation site on the transcript. During the second round of amplification and subsequent in vitro transcription, two goals are achieved:

1. Fluorescent label can be incorporated into the synthesized material.

2. Sufficient amount of this material is available for hybridization to the microarray.

Finally, it is necessary for the product of these reactions to be fragmented to allow an efficient and reproducible hybridization to the probes on the array. You may wonder why fragmentation is necessary, but the necessity emerges from the facts that the probes on the microarray are of a limited size and that the second round synthesis will faithfully reproduce the cDNA, regardless of length.

The earliest chips were '*spotted*' arrays, which consisted of a library of probes that were printed onto a glass slide using a head of needles. For these experiments, the control and the treated sample needed to be labelled differently (e.g. Cy3 or Cy5) and the samples combined prior to hybridization. Using different dyes carries the potential for bias (that the dye might affect hybridization efficiency), and this was controlled using a 'dye-swap' control that repeated the experiment, but with each sample incorporating the 'other' dye.

These early array experiments were plagued with technical difficulties, such as spotting irregularities (often called doughnuts, on account of their ring shaped misprint), and analytical challenges such as normalization strategies (see below). However, these pioneer chips became the basis upon which all current whole genome analysis experiments are now based. Modern microarrays eliminate many of the experimental problems that the spotted ancestors encountered, mainly because of the high throughput manufacturing with which the technology is produced. These developments bring the following advantages:

- Chip-to-chip variation is so small that control and treated samples can be hybridized to different microarrays.

- Separate chips are used for separate samples, a single label (biotin) can be used and the issues that surrounded dye-bias are eliminated.

- Probes are often synthesized directly from sequences that have been uploaded to public databases, so probe error is minimized and changes in gene annotation/function can be easily updated.

- Individual probes are often made of multiple overlapping smaller probes, so that some statistical inference of the level of non-specific hybridization can be drawn.

Regardless of the technology employed, microarray data relies on the relative hybridization ratio between the control sample and a treated sample to each single probe in the library on the chip. The result of a microarray experiment comparing two (or more) conditions is a list of probe identifiers that relate to genes, and expression values in which the end user has confidence are:

- expressed in the tissue/sample;

- differentially expressed by some prescribed fold-change cut-off criterion;

- significantly different between the treatments.

Having satisfied these criteria, this list of genes represents the minimum usage of the available data. More advanced bioinformatic analysis of such lists can establish gene-function, generate functional networks and drive hypothesis detection. It is important to note that while microarray *is* a hypothesis machine, it does not stand alone in experimental biology. The validation of some of the identified genes using an independent technique (e.g. qPCR – see Primer 20) will satisfy that the 'false discovery rate' (see below) is low and will give confidence in those other significantly regulated genes revealed in the experiment.

23.2 Required controls
23.2.1 Sample collection

Special care is required when collecting biological samples, extracting RNA and sample preparation, because RNA is very liable to enzymatic degradation by both endogenous and introduced ribonucleases (RNases). Fortunately, several commercially available chemicals ensure that RNases can be controlled and, together with experimental diligence, degradation can be kept at bay. All tools required for the dissection or culture of biological material under study must be free of RNase contamination, and any reagents required must be made using RNase-free solutions and chemicals and must be handled in an aseptic manner at all times. It is also appropriate to claim an area of the laboratory within which only RNA work will be performed (Figure 23.1a).

Once extracted, tissue or cells should be stored in an appropriate manner to protect against endogenous RNase degradation. Commercially available reagents that RNase-protect samples are available, but ultra-low temperatures help to ensure that samples are kept 'safe'.

Successful microarray experiments are dependent on the quality and quantity of the biological samples that are used. Experiments based on tissue dissections, are subject to huge sources of error that can result from including RNA from neighbouring tissues that either contaminate or dilute the biological signal being studied. It is therefore important for dissections to be maintained by a single competent operator, and for those dissections to be consistent between samples and based on appropriate anatomical atlas for that species.

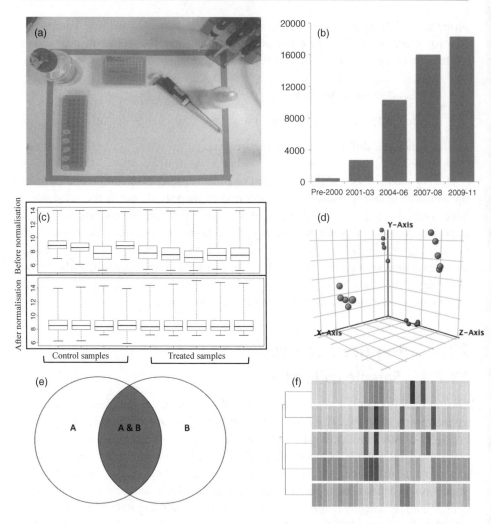

Figure 23.1 Overview of the features of microarray experiments. A: Meticulous attention to RNase control is required in microarray experiments. It is a good idea to establish a clean space within which RNase contamination can be controlled. B: The number of publications in the literature that use the term 'microarray' in the text has increased dramatically in the past 10 years. With over 47,000 publications, the quantity of data available is immense. C: Before and after normalisation of microarray data. In this box and whisker plot, each sample represents the spread of 31,099 data points and the bold bar represents the median expression value. Normalisation of microarrays ensures that variations in the data that are due to some technical aspect or some uncontrolled biological aspect are 'ironed out' between samples. D: Principle components analysis (PCA) establishes the degree of variability between samples; each point on this PCA represents a single sample and the expression of over 30,000 probes. Four groups of samples emerge from this analysis that correspond to three different treatments. E: The Venn diagram, useful for comparing different lists of regulated genes so that those commonly regulated elements may be distinguished from those that are unique to a particular condition. F. Clustering of gene expression data can act as a quality control check and identify gene expression patterns across multiple datasets; here the five groups are from the same brain region but the top two on the right cluster differently because they are from a different strain of animal than the other three. Courtesy of Dr. Charles Hindmarch.

RNA quality should be assessed following extraction to ensure that the material is of sufficient quantity and quality for use on the microarray. Technologies that rely on spectrophotometric analysis of RNA can establish the concentration, based on the absorbance at 260 nm (A260 = 1 = 40 µg/ml), and quality, based on the ratio between the absorbance at 260 nm and 280 nm (indicating contamination with proteins). RNA integrity (if the sample is degraded) can be assessed by running the sample on a denaturing gel or a microcapillary system that will show bands (or spikes) for the 18s and 28s ribosomal RNA in the samples whose integrity are a good proxy for mRNA quality.

Usually, the handling of RNA in an array experiment is boiled down to just a simple statement such as 'all dissections were performed in an RNase-free manner' or 'quality control was assured in each sample using spectrophotometric analysis'. It is not always the case that samples were collected and stored in laboratory conditions (collection of RNA in the field or biopsy following surgery, for example). In these cases, the integrity of the RNA might warrant specific reference to quality and consistency between samples.

23.2.2 Sample replication and sample pooling

The microarray experiment is not unlike any other; each measurement (each genome expression microarray) is a snapshot of a continuous biological process that varies according to treatment. To ensure that this snapshot is an appropriate one for the biological process, it is necessary to introduce both biological and technical replicates into a microarray experiment (see also Primer 2). These replicates are required to account for [2]*variations in the biology* and [3]*technical aspects of the experiment* that cannot adequately be controlled for within the experiment.

While both technical and biological replication can be achieved by increasing the number of arrays used in the experiment, pooling of sample from different animals onto a single microarray is a good way to reduce population level effects on the experiment. When using an outbred animal population, for example, each array can be turned into a microcosm of the animal population. When pooling, each array should represent samples that are independent from one another. The methodology of the paper should explicitly state the number of microarrays that have been used for each condition and the exact nature of the biological tissue that has been hybridized onto each array. Such information is a requirement on the public databases to which most journals require microarray data to be submitted.

---

[2] 'Variations in the biology' include circadian rhythm, oestrous cycling, outbred strain, etc.
[3] 'Technical aspects of the experiment' include experimenter variation, surgical precision and other such errors.

23.2.3   Normalization

Microarray data walks a fine line between the [4]*false positive* result and the [5]*false negative* result. Because microarrays represent so many probes, it can often be difficult to work out how to handle the data so as to avoid false positive results without being so stringent that you introduce false negative results. Careful normalization and statistical testing with appropriate *multiple test correction* (see below) is critical for minimizing false results in array experiments.

The process of normalization ensures that any technical (e.g. background signal between two arrays) or uncontrolled (e.g. different experimental batch) aspect of the experiment is removed from the data, so that the differences that remain result from the treatment under study. Figure 23.1C shows the expression value of nine microarrays; four independent control replicates and five independent treated replicates. The plot for each sample shows the range of expression values and the bold horizontal line shows the median value of each array. Clearly, the median value is more variable in the control state, and a difference between the medians of the control and the treated arrays appears to exist.

Two assumptions about microarray experiments need to be made in order to understand the reasons for normalization:

- First, it is expected that all arrays within the same condition *should* be broadly similar because they are replicates;

- Second, we expect the majority of the 31,099 probes on the array *not* to change, even in response to a treatment.

With these assumptions made, it is generally accepted that variations in the total expression range must come from non-controlled factors. Normalization nudges all the microarrays onto the same playing field so that these experimental artefacts do not overemphasize the expression of a gene (*false positive result*) or, indeed, mask it (*false negative result*). There are several different normalization strategies available, and their choice depends on various factors, including sample size and data quality. These should be outlined in the methodology of the microarray paper.

23.2.4   Multiple testing correction

In any experiment in which differences between one or more treatments and a control are sought, it is necessary to decide whether observed differences are

---

[4] **False positive**: the data says the mRNA *is* up-regulated when it *is not*.
[5] **False negative**: the data says the mRNA *is not* regulated when it *is*.

likely to be due to real differences among the groups (see both Primers 2 and 3). Statistical testing allows the following question to be answered; is the difference between two sets of data a product of chance or is it significantly different? If the probability that the result occurred by chance is less than five per cent ($p < 0.05$), then we can be 95 per cent confident that the difference is real. This five per cent can be considered a *false discovery rate* (FDR). A problem exists, however, when more than one test is being performed, because the false discovery rate changes with the number of tests, according the equation:

$$FDR = p\text{-}value\ cut\text{-}off \times number\ of\ tests$$

The result is that when tens of thousands of tests are performed as with an array experiment, the overall FDR will be near 100 per cent and there will be no confidence in any result. *Multiple test correction* is a statistical technique that accounts for the changing FDR by modifying the p-value threshold in proportion to the number of tests being performed, with the result that the *corrected* false discovery rate is always below 5 per cent. Most of the multiple test corrections rank the observed p-values of each test performed and correct as a function of the total number of tests. The main difference between the various correction protocols available is their stringency; given the high number of tests performed in an array experiment, even the seemingly smallest p-value can be rendered insignificant and can produce a false negative result.

23.2.5 Microarray data

Each probe on a microarray is assigned a unique identifier to distinguish it from all other probes and to provide a reference, so that they can be updated with changing descriptions and annotation. The basic information usually desired includes, but is not restricted to:

- Internal ID – unique identifier of probe assigned by the manufacturer
- Gene symbol
- Gene Description
- Genbank ID
- Unigene cluster ID

This is complemented by information gained from the experiment, such as:

- average raw expression value in a particular condition;
- p-values of probe;

- fold-change regulation between particular conditions;

- statistics involved in selecting differentially regulated genes.

23.2.6   Visualizing microarray data

Reading gene lists in journal articles is very boring – fact! It is a real challenge for authors to find interesting ways both to share the findings of their data and to engage with the scientific community about their meaning; figures in journal articles should be designed to convey those patterns evident in the data. Microarray data is usually presented in a few simple formats that have almost become as recognizable and easy to interpret to the lay scientist as a Western blot or a histogram.

The *principal component analysis* (PCA) allows the expression profile of all probes from each sample to be used to plot the relationship between the samples in a three-dimensional space. Each 'component' allows the degree of variation in a particular direction to be plotted, with the first component being the largest degree of separation, and each subsequent component representing smaller separations. Figure 23.1d represents 20 microarrays with four different treatments under investigation, and the PCA is performed on all 31,099 probes on each array. It is satisfying to see that the gene expression profiles correlate to treatment. The PCA is also an excellent method to identify outlier microarrays whose primary quality control should be checked to ensure that the array experiment was a success.

The *heat map* is a way to show the expression change of individual genes between two conditions using blocks of colour, but it is still just a gene list, albeit nicely coloured, so it has a limit on the number of genes that the reader can digest. A heat map becomes more useful when combined with a second analysis, such as *clustering* (Figure 23.1f). Such analysis will allow the author to show you that some conditions are more similar to each other than are others. For example, Figure 23.1f shows five conditions that have clustered according to the strain of animal used, suggesting that gene expression is highly strain-dependent (Hindmarch *et al.*, 2007).

Using conventional graphs can also be useful, but again only for relatively small numbers of genes, otherwise they lose their usefulness. The Venn diagram (Figure 23.1e) allows different lists of significantly regulated genes to be compared, so that both the number and proportion of genes that overlap between the two experiments can be visualized. In Figure 23.1e, a list of genes that are regulated by treatment A is compared to a list of genes that is affected by treatment B. The intersection between these two represents a list of genes that are regulated by both treatments. The sections that do not intersect, therefore, list genes that are affected by either treatment A or B. These lists now start to become more useful in downstream and more advance analysis that, for example, may seek to establish the function of those genes affected only by treatment B.

23.2.7   Advanced analysis

The advances in microarray experiments have certainly increased the amount of expression information available, but these data are being been generated at a faster rate than can be understood. The number of experiments that include microarrays has been increasing sharply year-on-year since 2000. At the time this primer was written (April 2012), the term 'microarray' returned over 47,000 hits on the website Pubmed (Figure 23.1B; www.ncbi.nlm.nih.gov/sites/entrez), with each publication potentially reporting the expression of hundreds or thousands of probes in a particular paradigm.

The non-biased approach to data collection is essentially undermined by an inadequate and heavily biased approach to investigation and analysis; scientists are still resorting to studying the genes on the microarray that they know! In order to investigate further the physiological functions of the genes regulated in an experiment, a Gene Ontology analysis can be performed. Each gene is annotated with various functional ontologies called GO terms, and such terms are split into three broad domains:

- Cellular component.

- Molecular function.

- Biological Process.

Within these domains, the vocabulary is well controlled and subject to constant flux in the light of novel research. GO analysis relies on the probability that a particular term will appear in any given gene list above that of pure chance (on the entire data set – see Hindmarch *et al.*, 2011).

Searching for patterns in vast fields of data that may convey meaning is a substantial challenge and has led to the development of novel approaches that have become associated with the term *bioinformatics*. The novel analytical approaches that characterize bioinformatics are ongoing and require the collaboration of biologists, mathematicians and computer scientists. Using these new methods, the complexity of life is being modelled, tested and re-modelled. Network reconstruction strategies are starting to make patterns from transcriptomic (and proteomic) data and are attempting to identify those genes in a list that are most important for the stability of the network and, thus, which are best to target in order to modify and manipulate a network and therefore, perhaps, the disease state that the network represents.

23.2.8   Experimental pipeline

The pipeline presented here is highly simplified and every stage requires careful optimization prior to the 'final experiment' being performed. Microarray experiments can be expensive, so care and attention should be lavished on each of the following steps:

1. Establish a biological question and design a robust experiment that will incorporate both biological and technical replicates.

2. Dissect/obtain the biological material from the experimental groups in the study and store in a manner that will reduce the risk of RNase degradation. *Be prepared to validate RNA stability.*

3. Extract and purify the RNA from these samples in a single batch (if possible) to reduce unnecessary experimental derived variability.

4. Selectively amplify the mRNA population from the total RNA population and use this as a template for both amplification and label incorporation. Fragmentation of this material should precede hybridization to the microarray.

5. Perform post-hybridization washing to ensure low signal-to-background ratio. Scanning images of array will allow feature extraction and production of raw data.

6. Inspect data quality and conduct normalization of the arrays in the experiment, so that artefacts such as background signal differences between arrays do not affect the false discovery rate.

7. Apply a statistical test to establish whether, for each probe, the expression levels are different between the control and the treated microarrays. Ensure that a multiple test correction has been applied that will modify the p-value threshold in proportion to the number of tests being performed, so as to ensure that the corrected false discovery rate is always below five per cent.

8. Consider how to visualize the result. Most often, this will involve using fold-change between the expression values so that the data represents both likelihood and magnitude of mRNA expression change as a consequence of the experiment.

9. Apply advanced bioinformatic techniques, including **gene set enrichment-analysis** (GSEA), biological pathways analysis and gene network reconstruction.

10. Validate the findings of the microarray in the biological question and try to alter the expression of the gene, in an attempt to establish physiological function of that gene in the system of interest.

23.3   Common problems or errors in literature

- As with any experiment, a good number of replicates are critical to the inference of the data! A minimum for a microarray experiment is $n = 3$ per condition.

- Ensure that microarray experiments have been normalized. Without appropriate normalization, the data may represent experimental noise.

- If a multiple test correction has not been applied, then scepticism should be applied with prejudice (Why was no multiple test correction been applied? Is there a problem with the experiment?). Multiple test correction can be very stringent and can result in a failure to identify [6]*significant genes*, and authors might have good reasons why they did not/could not perform the test. These reasons should be stated in the text; if they are not, then ask why.

- Raw and processed data should always be published on a public database such as the *Gene Expression Omnibus* (referenced below). This database collects all information about how the experiment was performed and what normalization/analysis strategy was adopted.

- Sometimes, a microarray will not include a '[7]favourite' gene (or the favourite gene of the boss) in the library. This is a disadvantage of microarrays, rather than a failing of the experiment – the microarray library is fixed prior to hybridization, unlike *next generation sequencing* (NGS: see below). Besides, finding a gene already known to be important in a biological system is just validation for a microarray; the real magic of transcriptomics is the identification of genes that no one would have thought to look for.

23.4   Complementary and/or adjunct techniques

Recently, *next generation sequencing* (NGS) has become the new star on the transcriptomics stage. Rather than making a library on the chip (as with microarray), the biologist makes the biological sample into the library and then literally grows this library onto the chip. This approach has several advantages over microarrays: the results are not limited by which probes were chosen for the arrays; and novel splice variants and single nucleotide polymorphisms (SNIP; pronounced 'snip') can be detected, as can insertions and deletions. These benefits must be weighed against cost and expertise required to analyze such complex data.

Further reading, resources and references

RNase control: http://www.invitrogen.com/site/us/en/home/References/Ambion-Tech-Support/nuclease-enzymes/general-articles/the-basics-rnase-control.html

Gene expression omnibus (public database for array data): http://www.ncbi.nlm.nih.gov/geo/

---

[6] **Significant genes**: defined in this sense as genes whose expression is changed more than the acceptance criteria.

[7] **'Favourite'** in this sense is taken to mean a 'likely candidate' – a gene that may be suspected *a priori* for a variety of reasons.

Affymetrix rat 230 2.0 Genechip: http://media.affymetrix.com/support/technical/datasheets/rat230_2_datasheet.pdf

Free analysis packages for microarray using the programming language 'R': http://www.bioconductor.org/

Hindmarch, C.C., Fry, M., Smith, P.M., Yao, S.T., Hazell, G.G., Lolait, S.J., Paton, J.F., Ferguson, A.V. & Murphy, D. (2011). The transcriptome of the medullary area postrema: the thirsty rat, the hungry rat and the hypertensive rat. *Experimental Physiology* **96**(5), 495–504.

Hindmarch, C., Yao, S., Hesketh, S., Jessop, D., Harbuz, M., Paton, J. & Murphy, D. (2007). The transcriptome of the rat hypothalamo-neurohypophyseal system is highly strain-dependent. *Journal of Neuroendocrinology* **19**(12), 1009–1012.

24
Experimental Proteomics

Thierry Le Bihan

School of Biological Sciences, University of Edinburgh, UK

24.1 Basic 'how-to-do' and 'why-do' section
24.1.1 What is proteomics used for?

Proteomics can be defined as the analysis of:

1. how much of a given protein or proteins is expressed (under specific conditions);

2. how proteins interact with each other; and

3. what is the nature and the dynamics of their modification (post-translational modification)?

Although it is often considered that the proteome is to proteins what the genome is to genes, several major divergences emerge, due to the differences both in their nature and in their respective roles.

24.1.2 How do proteomics and genomics differ?

A major difference between genomics (see Primer 23) and proteomics is that there is no protein equivalent to the polymerase chain reaction used for DNA amplification (PCR; Primer 20). Moreover, cellular proteins have expression levels that vary over a very large dynamic range (i.e. very low to very high levels), which is a significant challenge to proteomic techniques. Essentially, the dynamic range of a mass spectrometer must be able to cover at least four to five orders of magnitude in protein abundance, knowing that, within a simple cell like yeast, this difference can reach more than six orders of magnitude. At the extreme, this can mean attempting

Essential Guide to Reading Biomedical Papers: Recognising and Interpreting Best Practice, First Edition.
Edited by Phil Langton.
© 2013 by John Wiley & Sons, Ltd. Published 2013 by John Wiley & Sons, Ltd.

to detect a single molecule (copy) of one protein in a cell that may contain over a million copies of several other proteins and smaller quantities or many hundreds of others.

This difference, between what can be technically achieved in terms of detection and what is actually present within a sample, is often referred to in the literature as being simply the 'tip of the iceberg'. Having such a vast dynamic range to be covered is probably one of the most important challenges to tackle in the proteomics field, and it ultimately defines the chosen experimental design. In addition, proteins are continuously being synthesized and degraded, which adds the important variable of '*time*' to the protein detection equation. This characteristic is not captured at the mRNA level and explains why, in some cases, a poor correlation between mRNA and protein level exists.

This last example clearly supports the notion that for protein quantitation, it is better to use a direct approach (i.e. protein measurement) instead of an indirect approach based on mRNA transcript measurement.

Although proteomics does have its limitations, it has some advantages over other techniques as well. For example, it provides the ability to acquire, relatively quickly, either a global or a specific snapshot of the protein composition within a sample. This 'snapshot' can be obtained either with or without prior knowledge of cellular or tissue-specific protein abundance.

To some extent, some of the above-mentioned challenges in proteomics have been tackled by significant improvements in sample preparation and separation techniques and by improving the mass spectrometer instrument itself, as well as improved data analysis methods and bioinformatics.

24.2   Important considerations

24.2.1   Sample preparation and separation techniques

A typical proteomic mass spectrometry-based analysis is roughly based on protein or peptide separation, followed by analysis using a mass spectrometer with the peptide sequence often being confirmed from an already available database of known protein sequences.

In a typical proteomic approach (often referred to as 'bottom-up'), the protein samples are digested with a protease such as trypsin. The peptides extracted are then analyzed using a mass spectrometer, where peptide masses are measured (in MS mode) and using isolation and collisional activation energy in tandem MS (often described as MSMS or MS2).

The peptide sequence can be deduced from the mass/charge of its different fragments. Although a bottom-up approach is appropriate for protein identification in complete mixtures, not all the peptides will be identified in MS mode. Further-more, not all of the peptide fragments generated by ionization will be detected by the mass spectrometer, due to some of the physico-chemical properties of the peptide itself. Therefore, the incomplete protein coverage which is typical of a bottom-up approach often leaves gaps in the information obtained for a given protein (e.g. information may be missing regarding a protein's post-translational modification, splice isoform, specific site mutations). In addition, information about the mature protein sequence or its potential degradation may not be captured. All of these pieces of missing information can be important elements of a protein's biological function.

In most cases, bottom-up approaches fall into two main categories, both of which are designed to reduce the complexity of the sample being analyzed:

1. *Protein separation by **2-D**imensional gel-electrophoresis (2DE):* Protein mixtures are first separated in two orthogonal dimensions. Typically, proteins are separated based on their net charge (their isoelectric point). Next they are transferred to a SDS-PAGE gel (see Primer 15 for more detail), where they are separated again, this time based on their molecular weight. Although this is both a robust and valuable approach, it is being slowly excluded from the 'mainstream' proteomic field as other techniques become more popular. Nevertheless, the 2DE approach is often used in cases where an organism's genome has not been fully sequenced (e.g. in the fields of agronomy and environmental sciences). Following separation on gel, the proteins are visualized by the use of specific protein stains. The gel area containing a given protein is then excised from the rest of the gel, and the gel spot is subjected to proteolytic digestion and the peptides analysed by MS. Even though 2DE is a well-established approach, it is characterized by a very limited dynamic range and often poor transfer of hydrophobic membrane proteins from one dimension to the other.

2. *Multidimensional chromatography:* Another approach, developed in order to tackle the complexity challenge is the **multi**dimensional **p**rotein **i**dentification **t**echnology (MudPIT). This approach is based on the protease digestion of the entire protein extract. This is separated first on a strong cation-exchange column and, subsequently, each fraction is separated by reversed-phase chromatography coupled to mass spectrometry. The MudPIT approach is now considered to be a minimum standard for any shotgun proteomic study. In studies where each sample is fractionated to generate between 3–24 fractions, each of those fractions can take between 60–120 minutes on average to be analyzed on LC-MS. Therefore, each sample in a given shotgun proteomic study may require a complete day of LC-MS time to be analyzed. The ability of MudPIT to dig deeper into a given proteome comes at the cost of time invested into a given study

In addition to these main categories, there are specialist technical approaches that reflect a desire to understand a variety of specific modifications that are made to proteins and which have functional significance.

For example, *post- translational modifications* (PTMs) are the 'alteration' of specific amino acids within a protein sequence which change the properties of the protein. Protein structure, activity, function and stability are all influenced by various forms of PTMs. A few examples of post-translational modifications and their effects on proteins are:

- Acetylation for protein stability and regulation.

- S-nitrosylation and phosphorylation for signal transduction.

- Ubiquitination for proteolysis and protein sorting.

- Disulphide bond formation for stability or redox sensing.

Due to their low abundance and, in some cases, their instability, detecting any PTM using one of the previously mentioned methods mostly relies on chance and is very inefficient. A more pragmatic approach consists of enriching for the PTM in a sufficiently large quantity, which is only possible for few PTMs.

The various methods for PTM enrichment can be divided into several classes. They are:

1. chemical/physical affinity, such as immobilized metal affinity chromatography, which is based on coordinate binding between a metal ion (Fe, Ga) and a phosphate group on a peptide;

2. a depletion-based approach such as a CnBr column which, under specific conditions, will capture free N-terminal peptides (acetylated N-terminal peptides which do not bind to the column are found in the flow-through);

3. immunoaffinity purification techniques such as those that have been developed for ubiquitin and phosphotyrosine, for example;

4. other 'hybrid' or 'tandem' methods which are a combination of the above-mentioned methods.

A typical post-translational modification analysis could have the following pattern: PTM modified peptides are enriched using one of the methods described above and then analysed by mass spectrometry. The experimental peptide mass observed in MS mode are matched against theoretical potential peptides with and without mass difference associated to a given PTM. For a specific PTM, the modified amino acid remains unchanged (for example acetylation) and in MSMS mode the peptide fragments in the presence of a collisional gas (refers to collision-induced dissociation CID) in a similar manner as its unmodified counterpart except that a mass shift is observed for the modified amino acid.

Finally, in the case of peptide PTM enrichment, interpretation of the relative comparisons between samples can be ambiguous, as one cannot distinguish between variations in the levels of protein expression and the degree of PTM of a given protein.

24.2.2 The mass spectrometer: different ionization modes and instrument design

Ionization modes To this point, we have described how to reduce sample complexity (2D GE, MudPit) and how to ensure the sample is compatible with the MS analysis (protease digestion). In this section, we will describe the two main approaches used for protein/peptides ionization and a brief word on the different types of mass spectrometers. Proteins and peptides are non-volatile polar molecules requiring a soft ionization method for their transfer into a gaseous phase. The two main approaches used to achieve this are MALDI and ESI:

1. **M**atrix-**a**ssisted **l**aser **d**esorption **i**onization (MALDI). This involves mixing the sample with a matrix which absorbs laser energy that is transferred to the peptides. The laser heat simultaneously induces both the desorption of matrix and the transfer of singly positively charged ions of peptides into the gas phase. Some of the known drawbacks of this approach are the generation of single-charge ions that are often difficult to fragment for sequencing purposes, and variation of the signal intensity due to the sample preparation.

2. **E**lectrospray **i**onization (ESI). In this approach, ions are generated from a solution and are produced by applying a high voltage (2–6 kV) between the end of the solution separation device (commonly an HPLC column) and the inlet of the mass spectrometer. Under these conditions, an electrically charged droplet is created, which results in the formation and desolvation of analyte-solvent droplets. Under these ionization conditions, peptides often carry multiple charges and there is some interdependence between ion intensity, the ion concentration and the flow rate.

Mass spectrometers Specific types of proteomic applications are better suited to specific types of mass spectrometers. The main characteristics that differentiate by instrument design are:

- mass accuracy;

- resolving power;

- sensitivity (limit of detection, LOD);

- sampling rate;

- dynamic range.

These parameters are summarized in Figure 24.1. In general, for most proteomic applications, hybrid or tandem mass spectrometers are often used because information regarding the exact peptide mass needs to be extracted (obtained in MS mode), as well as the isolation and fragmentation which is performed in MSMS mode.

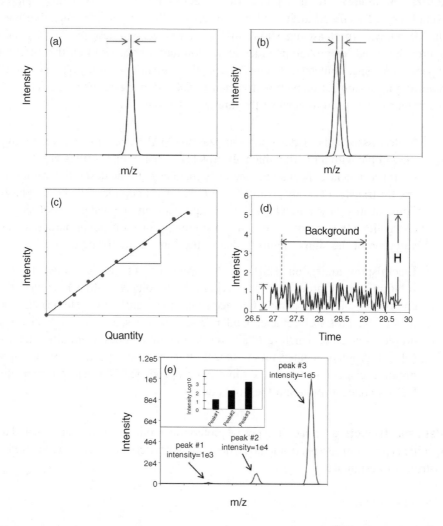

Figure 24.1 Visual representation of the different major characteristics that distinguish the different mass spectrometers. (a): the mass accuracy. (b): resolving power. (c and d): Sensitivity and limit of detection (LOD). (e): dynamic ranges. Courtesy of Dr. Thierry Lebihan. *A full colour version of this figure appears in the colour plate section.*

In general, MS instruments fall into three broad types:

1. *Time of flight* (TOF). These instruments are based on the principle that, for a given charge state, the time ions will take to reach the detector (under a potential) depends on, and is inversely proportional to, their mass (i.e. low mass ions will reach the detector before high mass ions). TOF instruments are often used in a hybrid configuration with a quadrupole (Q-Q-TOF), as well as in tandem (TOF-TOF), or even in a triple-TOF configuration. These instruments have a good mass accuracy and good quantitation capability. Furthermore, they are often used in discovery proteomics (Q-TOF), although the triple TOF from AB-Sciex, according to the vendor's claim, has been designed as a single platform instrument for both discovery and targeted quantitative proteomics.

2. *Quadrupole-based instruments.* A quadrupole is a structure composed of four parallel rods, where a radio frequency (RF) quadrupole field is generated which stabilizes the path of an ion having a given m/z ratio. This RF field can be adjusted incrementally, allowing the analysis of a wider m/z range. As for the TOF-based instrument, some of the quadrupole-based instruments are found in a combination of three quadrupoles, where the first and third 'quads' are mass filters and the second quadrupole is a collision chamber (ion fragmentation). These instruments have a lower mass accuracy and a lower resolving power than a TOF. Depending on how they are used, they can have a high dynamic range, and they are often used in targeted proteomics.

3. *Ion traps.* There are several types of ion traps. They are:

   • The *three-dimensional quadrupole* ion trap, which is based on the same principle as the quadrupole mass spectrometer described above. In this case, however, the ions are instead trapped and consecutively ejected.

   • The *linear ion trap* (LIT) differs slightly from the 3-D quadrupole ion trap because it is a two-dimensional instead of a three-dimensional quadrupole field. This allows the trapping of a higher number of ions (and increases the dynamic range of the instrument).

   • The *Orbitrap*, in which ions are trapped in an electrostatic field and rotate around a central electrode. Ions are characterized by two different oscillations; one is around the electrode, while the other is a back-and-forth movement along the electrode. The latter generates an image current which is dependent on the mass-to-charge ratios. As a result, Orbitrap instruments have high mass accuracy and sensitivity, as well as a better dynamic range, compared to the two previous ion traps.

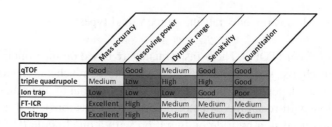

Figure 24.2   Main types of mass spectrometers and their characteristics. Courtesy of Dr. Thierry Lebihan. *A full colour version of this figure appears in the colour plate section.*

- *Fourier transform ion cyclotron resonance* instruments share some common features with the Orbitrap. In this case, the ions are trapped and oscillate within a magnetic field instead. Although they are more tedious instruments, they have high mass accuracy and sensitivity as well as a good dynamic range.

Figure 24.2 illustrates the strength of each type of mass spectrometer. None are perfect.

24.2.3   Issues of quantitation

Proteomics has moved from global analysis of organisms to the possibility of providing more information about proteins – for example, protein abundance, either in a relative manner (i.e. comparing control vs. disease state or drug treatment) or absolute quantitation. However, it is quite surprising to see quantitative measurements still being reported without any form of confidence in the measurements (either a standard deviation or a *p* value associated to the group comparisons). Quantitative proteomics often involves a sample preparation component as well as a bioinformatics one. In this section, we will concentrate on sample preparation.

Relative quantitation can be divided into three main groups:

1. *In vivo* metabolic labelling.

2. *In vitro* labelling.

3. Label-free.

Labelling refers to the use of a reagent, composed of light monoisotope (based on ^{12}C and ^{14}N) combined with stable heavy isotope (commonly based on ^{13}C and ^{15}N). Peptides from one sample in an experiment are labelled with light isotope, and peptides from another sample are labelled with heavy isotope. Both of the now-labelled peptide samples are mixed together and are similar enough to behave likewise for the overall procedure, whereas, for a given peptide, both forms (light and heavy) will separated at the mass detection level. The peak intensity ratio provides information about the relative abundance of the corresponding protein.

In vivo labelling strategies These can be performed by supplying cells with labelled amino acid (SILAC) or, in the case of autotrophic organism, using ^{15}N through nitrate or ammonium salt or ^{13}C (glucose, acetate and even CO_2). The **S**table **I**sotope **L**abelling by **A**mino acids in **C**ell culture (SILAC) has become a gold standard in the field of quantitative proteomics. Several amino acids can be used, but a 'classical' SILAC experiment is often based on a medium containing $^{13}C_6$-arginine and $^{13}C_6$-lysine. In this way, all tryptic peptides should have at least one labelled amino acid.

In vitro labelling strategies These are quite numerous in the field of mass spectrometry-based proteomics. One of the very first approaches developed was the **I**sotope-**C**oded **A**ffinity **T**ag (ICAT). This method is based on the modification of Cys-containing peptides with reagents of a different isotopic composition that yield a pair of ions 8 Da apart and subsequently to enrich for them by affinity. Several other *in vitro* labelling approaches have been developed, mostly targeting primary amine modification (which is the peptide N-terminal and the lysine side-chain).

Several issues with *in vitro* labelling have been identified, such as possible side reactions with amino acids other than the ones being targeted or incomplete reactions. Another main drawback of these labelling strategies is that they suffer from an increased sample complexity after the different samples have been mixed together (a reminder that complexity is the main challenge in the proteomic field). It is also difficult in some cases, if not impossible, to compare more than 2–8 samples at the same time. This imposes some constraints in terms of experimental design using these labelling strategies.

Label-free approaches A *label-free* differential approach compares LC-MS datasets based on relative peptide peak intensities, or by comparing the number of spectra acquired. A label-free quantitation strategy has fewer limitations in terms of the number of runs to compare. Using this method, either a few runs or up to one hundred runs can be technically compared.

Although label-free quantitation is a robust quantitative method, its performance depends on temporal LC-MS alignment of the different runs, which can be challenging. Since none of the samples being analyzed are encoded by isotopic labelling or mixed at any level, label-free quantitation is also strongly dependent on the reproducibility of the overall platform from the sample preparation to the LC-MS analysis. As more and more free label-free platforms are being made available and are priced in order to make them accessible to standard proteomics labs, this approach will undoubtedly increase in popularity. A comparison of the three main quantitative proteomics strategy and at which point samples can be combined is illustrated in Figure 24.3.

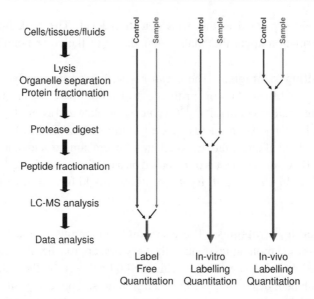

Figure 24.3 Typical quantitative mass spectrometry workflows. The left part is a generic process presented for a bottom up proteomics analysis. Mixed red and blue lines indicate when the two samples are normally combined together. The longer the process is run in parallel prior mixing the sample, the more technical variations are introduced, which can affect both samples independently. Courtesy of Dr. Thierry Lebihan. *A full colour version of this figure appears in the colour plate section.*

24.3   Required controls

24.3.1   Control definition

Depending on the type of experiments (immunoprecipitation or global proteomic survey), as well as the number of replicates and fractions per sample replicates, the type of control and the experimental design can vary significantly. For an immuno-precipitation experiment, where antibodies are attached to beads and protein complexes are captured, it is important to use well defined control (see Primers 13 and 14). The nature of the control can be, for example, using a non-specific antibody with the same type of samples as used for the experiments. If the effect of a perturbation is studied, then performing the immunoprecipitation enrichment both prior to the perturbation and after could highlight significant differences. As the performance of the mass spectrometer will vary with time, and some of the experiment could necessitate several days of mass spectrometry time, it can be important to randomize the samples to be run.

24.3.2   Sample normalization

Normalization has to happen at least at the following two steps:

1. Protein assay: to ensure that the same amount of sample is analyzed by LC-MS.

2. At the LC-MS trace: each sample signal output has to be normalized for proper quantitation.

In the next section, I discuss in more detail the importance of replicates and their nature.

24.4   Common problem or errors in literature and pitfalls in execution or interpretation

New and cutting-edge techniques in mass spectrometry are often limited to the inventor's research laboratories (and their close collaborators). Moreover, the newly introduced methods are often verified using rather simple models (e.g. BSA digests, or casein digests as a model for phospho-protein enrichment or analysis). In the best cases, the results obtained along these lines of experimentation should be considered simply as interesting proofs of principle. Extrapolating their performance to more realistic samples is erroneous and needs to be proven experimentally on relevant samples.

Even with the current improvements in instrumentation, recent MS-based techniques can only cover a fraction of a given proteome. Any small increase in proteome coverage is often accomplished by a concomitant increase in the amount of sample fractionation, which is a time-consuming operation. In these circumstances, the choice of digging deeper into a given proteome is often achieved at the expense of acquiring higher number of replicates. Ultimately, this has an impact on the conclusions drawn from the observed data.

How to define the number of replicates, as well as the type of replicates (technical versus biological), is important (see Primers 2 and 3). A technical replicate consists of repeating the analysis of the same sample several times. Such an approach allows for the evaluation of the variability inherent to the technique being applied. While this may be useful to extract, has it given some information about the technique used? Ideally, no biological interpretation should been drawn in these circumstances.

The definition of biological replicates in the proteomic context has not been adequately addressed as of yet, and is currently broadly defined. For example, a different flask of the same culture can be considered a biological replicate. Furthermore, different mice of the same genetic background are also considered to be biological replicates. It is essential to define the question needing to be answered by the proteomic study before deciding upon the experimental design. Only then can an appropriate experimental design and adequate sampling be performed in order to allow for the correct conclusions to be drawn (see Primers 2 and 3).

The higher the similarity between the different 'biological replicates' (i.e. simply different flasks of the same cells cultured in exactly the same way or same mouse genotype) will probably allow for the observation and reporting of nice, well-defined trends. However, the conclusions drawn should ideally be limited to the observed specific culture conditions. Using two different mouse genotypes also generates a wide range of different background proteins which, under certain circumstances, render data analysis almost impossible.

In the case of the number of replicates to consider in a study, triplicates are often arbitrarily considered 'safe'. However, a more robust approach, such as the use of a power analysis, which allows for the evaluation of an adequate sample size, should be considered. This step can be important, as the sample size may be either too low or potentially too high (rarely the case in the proteomic field). If too low, the experiment will ultimately lack the precision to give reliable answers in relation to the questions. On the other hand, if the sample size is too large, then resources and time will be unnecessarily lost, with little information gained. A power analysis can also be an extremely useful tool for determining ratio cut-offs between two groups being compared, instead of using the often employed and arbitrarily defined ratio cut-off of 1.5–2. In time, some standardization rules, not defined arbitrarily, will emerge in the field of quantitative proteomics. For example, proper power analysis should be mandatory in order to justify the chosen experimental design.

The reader would perhaps have appreciated a clear and definitive description of which methods should be used for mass spectrometry-based proteomic studies. However, different proteomic platforms provide solutions to different proteomic questions. Several major challenges still exist, while new ones are continuously emerging. Therefore, the overall state of the field of proteomics is far from being stagnant. In this primer, we haven't proposed a single solution; however, we have raised some key points to consider when reading or writing a proteomics research paper.

With time, a proteomic experiment will become easier to perform and more reproducible. Digging down into a proteome will hopefully become attainable, with a greater ease, and thereby allow researchers to explore other dimensions of biological significance, such as time series, several perturbations and a higher number of biological replicates. Under those circumstances, the field of proteomics will be able to realize its full potential and deliver on its tremendous promise.

24.5   Complementary techniques

A transcriptomic study offers several advantages over a proteomics one, as it is slightly easier to perform; often, more complex experimental design and number of replicates can be defined. However, as mentioned earlier, there are some time differences between the levels of protein and their corresponding mRNA. The transcriptomic approach is often done with prior knowledge, where a proteomics approach can be used as a non-hypothesis driven approach.

Protein array, based on antibody capture, is a technology which is developing fast and will soon be a serious competitor to the classical proteomic approaches; the expression and purification of diversified population of proteins will get easier, and there will be a larger bank of antibodies available.

Primer 13: Immunocytochemistry
Primer 14 Immunopreciptation
Primer 15: Immunoblotting
Primer 19: Viral vector transgenesis
Primer 26: Genetically modified models

Acknowledgements

SynthSys is a Centre for Integrative Systems Biology (CISB) funded by BBSRC and EPSRC, reference BB/D019621/1.

Further reading and resources

Several good reviews on the topic exist, as well as good web links.

Domon, B. & Aebersold, R. (2006). Mass spectrometry and protein analysis. *Science* **312**, 214–217.

Elliott, M.H., Smith, D.S., Parker, C.E. & Borchers, C. (2009). Current trends in quantitative proteomics. *Journal of Mass Spectrometry* **44**(12), 1637–1660.

Some tutorial on internet: http://www.i-mass.com/guide/tutorial.html

Some tools for proteomics analysis: http://expasy.org/proteomics

Section C

In vivo/Integrative

25
Behavioural Methodology

Emma Robinson
Physiology & Pharmacology, University of Bristol, UK

25.1   Basic 'how-to-do' and 'why-do' section

Behavioural methodologies are used to quantify specific behavioural outputs. The measurement of behaviour is a hugely diverse area, encompassing studies such as, the assessment of basic sensory or motor function, complex cognitive behaviours such as learning and memory, animal models of disease and phenotyping genetically modified animals. In this brief overview of behavioural methodology, issues relating to the following are all considered, and specific examples are used to illustrate key points:

- The choice and validity of a method.

- Animal models of disease.

- Experiment design.

- Data interpretation.

A detailed review of the different methods available is not provided. Further information relating to individual methodologies can be sourced by referring to literature databases or relevant textbooks and protocol journals (e.g. *Nature Protocols*). Examples of different behavioural methods and a brief synopsis are given in Tables 25.1 and 25.2 at the end of this primer.

25.2   Animal models and behavioural testing

In biomedical '*behavioural*' research, an *animal model* is a term used to describe experimental studies, using animals, that provide insight into human health and disease. In many cases, the animal models and clinical tests in humans measure very

Essential Guide to Reading Biomedical Papers: Recognising and Interpreting Best Practice, First Edition.
Edited by Phil Langton.
© 2013 by John Wiley & Sons, Ltd. Published 2013 by John Wiley & Sons, Ltd.

Table 25.1 Examples of different behavioural methods used to assess specific aspects of sensory or motor function.

Method	Primary application	Summary
Locomotor activity	Drug-induced sedation Control for methods which depend of motor function	Animals are placed in a test arena, where their total activity is quantified. This can be carried out using an automated system, with infrared detectors used to quantify both horizontal and vertical movements, or by manual counting.
Rotorod	Motor coordination/ ataxia	This uses a customised piece of equipment where animals are placed onto a rod, which is then turned at an increasing speed until the animal falls. Impairments in motor coordination and sedation are detected by a reduction in time on the apparatus.
Acoustic startle Pre-pulse inhibition (PPI)	Auditory response and sensory-motor gating	Animals are placed in an apparatus which detects vibration. A loud tone is played and the subsequent startle response is quantified. When combined with a pre-pulse (preceding low volume tone), normal animals show a reduced startle response. This PPI tests sensory motor gating.
Tail flick/hot plate test	Sensitivity to noxious stimuli Pain research	Animals are exposed to a non-damaging noxious stimuli and their time to respond to the stimulus is quantified. Impaired sensory function and antinociceptive treatments increase the time to detect and respond to the stimulus.
SHIRPA (**S**mithKline Beecham, **H**arwell, **I**mperial College, **R**oyal London Hospital, **p**henotype **a**ssessment)	Series of tests designed to assess the phenotype of genetically modified mice	This is a standardized set of procedures used test muscle function, cerebellar function, sensory function and basic neuropsychiatric function in genetically modified mice. Relevant to assessing phenotype, as well as determining potential confounds when using other behavioural methods.

similar processes (e.g. detection of a noxious heat stimulus). In other cases, the model has been validated (see below) in terms of its ability to replicate some aspect of human disease symptom(s) and/or to predict drugs that have efficacy in treating disease. For example, the forced swim test is an animal model of depression which can predict drugs with antidepressant efficacy; however, this model does not directly mirror any aspect of the human depression, nor the methods used to quantify mood in humans.

Table 25.2 Examples of different behavioural models associated with specific CNS diseases.

Method	Primary application	Brief description
Elevated plus maze	Anxiety research	Method uses rodents, natural aversion to height and open spaces. Rodents tend to spend more time exploring the safe 'closed' arms, but anxiolytic treatments increase time spent in the more aversive 'open' arms.
Forced swim test (tail suspension test)	Depression	Animals are normally exposed to the test environment (inescapable swim or suspension by the tail), then re-tested 24 hours later. Animals' normal response is to show reduced escape behaviour on re-exposure, but this is prevented if pre-treated with an antidepressant.
Novel object preference test	Recognition memory	Method to study short-term and long-term recognition learning and memory. Relies on rodents' natural predisposition to explore novel objects. Following exposure to two or more objects, animals are re-tested after a given period of time, with or without treatments, and memory for the familiar versus novel object assessed using a ratio of exploration time at each object.
Morris water maze	Spatial working memory	Method to study learning and memory in a spatial task. Animals are required to learn the location of a platform hidden just underwater. The time to learn the location and ability to remember the location are then tested.
Conditioned place preference	Addiction	Method used to study drugs with inherently rewarding properties. Association between an environment and drug or control treatments are made, and then animals preference-tested using a choice test where both environments are made available. Drugs which are reinforcing induce preference for the drug-paired environment. This method has also been used to look at conditioned aversion.
Pavlovian conditioning/autoshaping	Associative learning	Based on the initial work of Pavlov, this method is used to investigate the formation of an associative memory following pairing of a stimulus (e.g. light or tone), with an outcome (e.g. food reward or footshock). Following exposure to the stimulus, animals learn the association, which is quantified by measuring their approach or avoidance responses.
Rotational behaviour	Basal ganglia motor function (Parkinson's disease)	This method combines a neurotoxic lesion, 6-hydroxydopamine, and behavioural measure, to investigate motor function involving the basal ganglia. Following a unilateral lesion of the dopamine nigrostriatal pathway, animals exhibit rotational behaviours, particularly when an exogenous dopamine stimulus is used. Drugs with efficacy in PD attenuate this rotational behaviour.

Other behavioural methods are considered to be tests of a specific aspect of physiology (e.g. motor function, sensory processing). These tests are designed to assess the functional consequences of a given manipulation in terms of a behavioural endpoint. This is likely to involve the integration of a number of different systems of the body, and it takes great skill to ensure that the choice of method and the interpretation of the data are appropriate. While *in vivo* behavioural studies provide essential information about how different systems of the body integrate to mediate that response, non-specific effects can readily *confound* the results. This is particularly the case when looking at genetically modified animals, where a full behavioural screen is essential before any specific animal model of disease is used. The vast majority of behavioural tests and animal models rely on an animal's movement to provide the quantifiable measure. If the animal has impaired sensory or motor function, these are likely to confound the results.

Example 1   The results shown in Figure 25.1 illustrate data obtained from an experiment using the elevated plus maze, an animal model of anxiety. In this model, non-specific effects associated with both strain differences in baseline behaviour and locomotor effects are illustrated.

Example 2   If a genetically modified mouse has a locomotor impairment, it may appear to have impaired abilities in the Morris water maze test of spatial learning, but in fact it may just be unable to swim properly.

Example 3   If a genetically modified mouse has impaired hearing, it may appear to show attenuated learning in conditioned fear paradigm, but in fact it may simple be unable to hear the tone which predicts footshock and so fails to respond in the manner typical of an animal with normal hearing.

25.2.1   Validity

Behavioural tests are required to be '**valid**' (from the French '*valide*', meaning 'strong'). The validity of an animal model is tested against a number of criteria:

- How well the animal model reflects the symptoms in man is referred to as *face* validity.

- How well the animal model replicates underlying biology is referred to as *construct* validity.

- How well the animal model predicts treatments which will be effective in the human condition is referred to as *predictive* validity.

- How readily data obtained from the animal model can be translated to a benefit in the clinic is referred to *translational* validity.

Schematic representation of anxiety testing apparatus and data

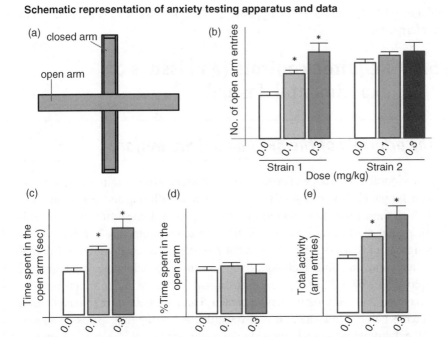

Figure 25.1   Schematic representation of anxiety testing apparatus and data. The elevated plus maze is one of the most widely used methods to study anxiety related behaviour in rodents. It utilizes their natural aversion to height and open spaces, and provides a number of different measures of anxiety-related behaviour and general activity. Panel A illustrates the basic apparatus. Two arms are open, while the other two arms are enclosed. The animal is usually placed in the centre and behaviour is recorded and analyzed over a period of time ($\approx$10 minutes). A key measure of anxiolytic behaviour is an increase in the time spent in the open arms. Panel B illustrates the type of data which may be obtained in this model. Note that there is a difference in baseline behaviour between the strains, and the response to the drug treatment differs. It is well known that different strains exhibit very different behavioural profiles, often exhibiting effects greater than the drug treatments. A common mistake when looking at EPM data is a failure to consider non-specific effects such as differences in motor function. The behaviour of the animal in this task is dependent on exploration, so an apparent decrease in time on the open arms (panel C) may appear to show an anxiogenic effect. However, analyzing these data as a relative value (i.e. percentage time in the open arm (panel D) reveals no overall difference. When the overall activity of the animals is analyzed (panel E), it is clear that the effects in panel D are due to an overall increase in locomotor activity. (*indicates a significant difference from control treated animals). Courtesy of Dr. Emma Robinson.

It is rare that an animal model can achieve all three levels of validity, particularly in relation to CNS diseases. Recently, emphasis has been put on translational validity, reflecting the commercial interest in biomedical research and drug development. Whilst designing an animal model which can achieve validity across all these areas is difficult, a lack of validity in one or more area does not necessarily restrict the use of the model, but it should be a key factor when deciding which model is most appropriate to the hypotheses being tested and how the data obtained are interpreted.

25.3   Required controls (and issues of experimental design)

The principle of the 3Rs and animal welfare

Behavioural studies depend on the use of a conscious living animal, so therefore there is potential for an animal to experience pain suffering and lasting harm. In the UK, these procedures will most likely require a Home Office licence, with other countries having similar national or local requirements. The experiment design should consider the '3Rs' – reduce, refine and replace – and good animal welfare and handling techniques are essential to achieving high quality, reproducible data. Stress responses in animals can have major effects on their behaviour, and these may even be greater than the effect of the manipulation which is being investigated. It is also worth considering that a stressed animal will exhibit physiological, endocrine and neurochemical changes, which will interact with the treatment administered.

This section is structured to provide some general information relating to the use of behavioural methodology. Some points may not be relevant or practical when considering a specific method. It may prove useful to review Primer 2 (on experimental design) and Primer 4 (statistics).

25.3.1   Behavioural testing equipment

The majority of behavioural testing carried out as part of drug development uses automated equipment. This reduces the time taken to complete a study, allows for more animals to be tested at a given time and reduces experimenter error and bias. Purpose-built automated equipment is expensive, so therefore is not necessarily realistic for smaller academic research groups. Not all methods can, or have been, automated, and many research groups still carry out non-automated behavioural testing.

25.3.2   Choice of method

It is essential that the experiment design is based around a central hypothesis. The behavioural method or methods are then chosen, based on the purpose of the experiment. Consideration should also be given to the specificity of the test to the behaviour of interest and whether additional, control tests should be included (e.g. assessment of motor function). Having chosen the method(s), consideration should be given to the limitations of the method and their impact on the interpretation of any data. A common problem with behavioural studies is the over-interpretation of a specific behavioural deficit – particularly one that has been observed but which is not consistent with the hypothesis.

25.3.3   Experiment design

The design of any experiment is vital to ensuring the reliability of the results obtained. With behavioural methods, this is particularly important, as many factors can lead to experiment design bias or experimenter bias (see Primers 2 and 3). Key design issues are:

- *Avoiding experimenter bias.* It is essential that any non-automated behavioural testing is carried out with the experimenter blind to treatment, otherwise experimenter bias will undoubtedly influence results (this is probably true of any experiment, but behavioural studies largely follow this procedure).

- *Counter-balancing study design.* Behavioural experiments can use either *between-subject* designs (animal only receives one treatment and observations are regarded to be unpaired) or *within-subject* designs (each animal receives all treatments and observations are regarded to be paired). A between-subject design avoids adaptation to repeated testing in the apparatus (e.g. elevated plus maze), but introduces a larger scatter (variance) within the data, which can require larger numbers of animals in each group. Within-subject studies tend to reduce variance and the numbers of animals needed, but are only suitable where animals are unlikely to adapt to the test or where the animal has been trained to a stable level of performance. Within-subject studies should provide statistical justification for this. Whichever type of experiment design is used, treatments must be fully counterbalanced to avoid bias through experiment design. This is normally done using a technique known as a '*fully randomized Latin Square*' design. This means that on any given day, all treatments are represented and factors such as the time of day are balanced within the study.

- *Species, strain and genetic modification.* Different species offer advantages and disadvantages, depending on the method and experimental objectives. Choosing the right species and strain for the behavioural method can greatly increase the validity of the data obtained. For example, the forced swim test (FST) was originally designed for rats, which are a species known to swim within their natural environment. In contrast, mice are highly averse to water and experience a significant stress response when exposed to the forced swim test, which can confound the results. As an alternative, the mouse tail suspension test was developed to apply the same principles as the FST, but using a more species-relevant procedure.

- Different strains of animal also show a high degree of variability in terms of their behavioural responses, and this can be both an advantage and a disadvantage. For example, some strains show higher baseline anxiety and locomotor activity in a novel environment. These strain differences can been used to facilitate detection of specific treatment effects, but can also result in *false positives*. For example, if the aim is to study the efficacy of an anxiolytic drug, selecting a strain of animal with high baseline anxiety will increase the chances of detecting an effect (see Figure 25.1, panel B). In contrast, strain differences in baseline performance in a number of behavioural tasks have led to over-interpretation of data from genetically modified mice where strain-matched controls were not used.

- *Control experiments for non-specific effects.* The majority of behavioural methods depend on animals using some sensory and/or motor processes to perform the test. If the manipulation used causes a generalized impairment in sensory or motor function, this may lead to an apparent deficit in the test when, in fact, it is a non-specific effect. In order to control for non-specific effects, the inclusion of additional behavioural tests may be necessary. This is particularly relevant when looking at the behavioural consequence of genetic modification. It is also important when assessing the effects of a novel pharmacological agent where effects on locomotor function, for example, have not been established. As most behavioural methods depend on motor function, tests of motor coordination and sedative effects are useful. If the test utilizes a specific sensory domain (e.g. visual processing), some assessment of normal function should be included when using genetically modified animals. For most laboratory strains, information about sensory functions is already known.

- *Combining behavioural and pharmacological methods.* The most common types of experiments performed using behavioural methods are pharmacological studies, where animals are treated with doses of a drug and the behavioural effects are quantified and compared to vehicle controls. Antagonist experiments are also performed using behavioural methods in order to assess the receptor(s) involved in mediating the response elicited by a given agonist. A typical agonist dose response experiment and antagonist study is illustrated in Figure 25.2. In this example, animals receive different doses of the agonists in order to establish the dose response relationship and EC_{50}. These experiments often generate bell-shaped dose-response curves, where the quantified behaviour initially increases (or decreases), and then non-specific effects start to counteract the effects or inhibit the animals' ability to express the behaviour (e.g. induction of sedation). In order to test whether this effect is mediating by a specific receptor, a second experiment is carried out, where a selective antagonist for the hypothesized receptor is used. In this second study, an EC_{50} dose of the agonist is normally used, and this is ideally tested against two doses of the antagonist. Although not always feasible, the doses of antagonist should not affect the behaviour when administered alone, and specifically and dose-dependently attenuate the agonist-induced behavioural response.

In vivo dose response data

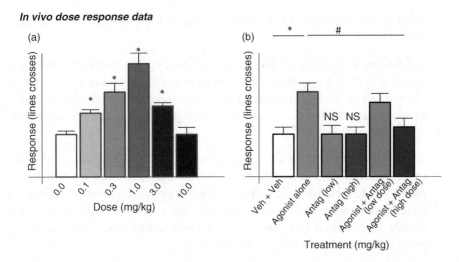

Figure 25.2   *In vivo* dose response data. These two figures provide an example of the type of data which may be obtained from a behavioural experiment looking at locomotor activity. The results in panel (a) show the dose dependent effects of an agonist. There is an increase in locomotor activity exhibited over a relatively small dose range (1 log unit), with an overall bell-shaped dose response effect observed. In many cases, the behavioural method is only sensitive to effects over a narrow range because low doses are sub-threshold and higher doses induce other, non-specific effects, which themselves affect behaviour. For example, the drug used in this experiment may activate receptors which increase locomotor activity at low doses (specific effect) but, at higher doses, other receptors are activated or inhibited, leading to non-specific effects. The data shown in panel (b) illustrate a well-designed experiment to investigate the receptor involved in mediating a behavioural response. An antagonist selective for the hypothesized receptor is used in the study. A total of six treatment groups are included to provide controls. The vehicle group is used to control for the procedure and vehicle used to dissolve the drug. The agonist-alone group is a positive control to ensure the effect is present. The antagonist-alone groups are included to test whether the antagonist has any effects when given alone at the doses to be tested in combination with the agonist. The agonist is then also tested in the presence of two antagonist concentrations. Ideally (although often not realistic), the antagonist alone has no effect and a dose-dependent attenuation of the agonist effect is observed. (*indicates a significant difference from control treated animals, #indicates a significant difference from agonist alone treated animals). Courtesy of Dr. Emma Robinson.

25.4   Pitfalls in execution or interpretation

The points below can be summarized as issues that require consideration when interpreting behavioural data and translating the findings to human disease:

1. How relevant is the behaviour that has been measured in the animal to the human symptom or disease?

2. Is the deficit or improvement observed specific? Have control behavioural tests been included or referenced?

3. Is the experiment design unbiased?

4. What is the specificity of the drug(s) used in the experiment at the doses administered?

5. Are the doses of drug used relevant to clinical doses? How does the dose used compare to the occupancy of the receptors *in vivo*? (This is not always easy to find out, but very relevant; different species metabolize drugs at different rates, so comparing receptor occupancy is the most reliable way of knowing that the dose is relevant.)

6. What is the time course of the drug effects and are these relevant to the experiment design? For example, targeted brain infusions may reach receptors after five minutes, while oral drug dosing may require one or two hours for peak plasma concentration to be reached. When using an antagonist, this is ideally administered first, and at a time which allows it to equilibrate at the receptor before the agonist is administered.

Further reading and resources

Crawley, J.N. (1999). Behavioral phenotyping of transgenic and knockout mice: experimental design and evaluation of general health, sensory functions, motor abilities, and specific behavioral tests. *Brain Research* **835**(1), 18–26.

Hatcher, J.P., Jones, D.N., Rogers, D.C., Hatcher, P.D., Reavill, C., Hagan, J.J. & Hunter, A.J. (2001). Development of SHIRPA to characterise the phenotype of gene-targeted mice. *Behavioural Brain Research* **125**(1–2), 43–47.

Nestler, E.J. & Hyman, S.E. (2010) Animal models of neuropsychiatric disorders. *Nature Neuroscience* **13**, 1161–1169.

26
Genetically Modified Mouse Models

Nina Balthasar

Physiology & Pharmacology, University of Bristol, UK

26.1   Basic 'how-to-do' and 'why-do' section
26.1.1   Why?

In the post-genomic era, we now have an enormous amount of information on the human genetic code but, of the estimated 21,000 protein-encoding genes, we still only know the function of roughly half. In addition, knowledge of the human genetic code is now allowing us to search for genomic variations across the population. Significant advances in large-scale searches for genetic determinants of human disease, using analysis of data from multiple genome-wide association studies, are helping to identify more and more loci and genes involved in particular diseases (see Primer 23).

However, in most cases, we still need to understand how these genes might fit into the pathways leading to disease. To investigate the physiological function of these novel genes, one can manipulate their expression (increasing or decreasing it) in mouse models and study the subsequent alteration of mouse physiology. It is thought that only about one per cent of human genes do not have a mouse counterpart. *In vivo*, whole-animal studies are particularly important – and actually irreplaceable – when multiple tissues are involved (e.g. when a gene in one tissue exerts effects on other tissues). A good example is the study of glucose homeostasis, in which the brain senses the glycemic state of the body and is able to adjust blood glucose levels by affecting, for example, glucagon release from the pancreas.

Essential Guide to Reading Biomedical Papers: Recognising and Interpreting Best Practice, First Edition. Edited by Phil Langton.
© 2013 by John Wiley & Sons, Ltd. Published 2013 by John Wiley & Sons, Ltd.

26.1.2   How?

There are two main techniques to genetically manipulate mice:

1. *Transgenic* techniques.

2. *ES cell* techniques.

Each of these has its own applications, advantages and disadvantages. Importantly, the term 'transgenesis' is often misused, by text books and journal articles alike, to encompass all forms of genetic manipulation in mice. Strictly speaking, it should only be used to describe transgenic techniques, as explained below.

Transgenic techniques   Transgenic techniques are often used in order to express a 'marker' (e.g. green fluorescent protein – see Primer 16) or to over-express a gene or a dominant-negative version of a gene in a very specific cell type (see Primer 22). Transgenesis basically involves the *random* insertion of an 'extra' piece of DNA (the 'transgene construct') into the mouse genome.

To target the marker or gene of interest to a specific cell type, it is necessary to drive the expression using a gene promoter that is usually expressed in the cell type of interest – and, ideally, only in that cell type. One thus generates a transgene construct by linking the cDNA of the marker/gene of interest to the relevant promoter.

If, for example, the intention were to express eGFP in all cholinergic neurons, one would link eGFP cDNA to a promoter that drives expression of a gene that is more or less specific to cholinergic neurons, such as the promoter of the choline acetyl transferase gene. The DNA transgene construct, consisting of the promoter plus the cDNA of interest, is then injected into fertilized mouse eggs and is integrated randomly into the genome of the fertilized egg as the genome replicates for cell division. The eggs are then implanted into a surrogate female mouse, which gives birth to 'transgenic' pups, if the integration of the construct was successful. Each transgenic pup originating from a single injected egg will, at this stage, have the construct integrated into a different locus in the genome; we call these transgenic pups 'founders' of a certain 'line'. Once we breed these transgenic founders, all offspring from the same line will have construct integration in the same genomic location.

It is important to remember that, in transgenesis, an extra, 'standalone' piece of DNA is added to the genome. Endogenous genes have not been altered as such.

Transgenic techniques are often favoured as they are faster and cheaper than ES cell techniques.

Es cell techniques   Embryonic stem cell (ES) techniques are used to manipulate endogenous genes or gene areas. They are *targeted* manipulations of specific, chosen areas of genomic DNA. They are often used to delete a gene (knock-out mice) or to insert mutations into a gene (knock-in mice).

To alter a gene functionally, it is still necessary to generate a construct – this time using genomic sequences of the gene (not cDNA) – and insert, for example, a *neo* cassette (a piece of DNA expressing the antibiotic resistance gene for neomycin) in a place where it disrupts the correct expression of the gene of interest. These sequences are then inserted into ES cells, which are 'pluripotent' (meaning they can develop into any kind of cell type), by electroporation.

ES cells use a process called 'homologous recombination' (exchange of DNA sequences to increase genetic variation) and will switch the endogenous sequences of the gene for the ones harboured by the construct inserted by electroporation. Cells that have undergone homologous recombination will be resistant to neomycin and can be selected for on that basis. These cells are then injected into the blastocyst stage of a fertilized egg and implanted into a surrogate female mouse. The developing embryo will thus have some cells derived from the original blastocyst and some cells derived from the introduced ES cells.

The hair colour gene encoded by the blastocyst is chosen to be different from that encoded by the ES cells (by using blastocysts and ES cells from different mouse strains). This allows us to identify offspring with most of their genome (and thus, hopefully, their germ cells) derived from the ES cells by a high percentage of 'ES cell hair colour' (such an offspring is called a high percentage 'chimera'). These high percentage chimerae are then bred to verify germline transmission of the altered gene.

Combination of transgenic and ES cell techniques in the Cre/loxP system The 'Cre/loxP system' is a technique that is often used to delete a gene [product] from a specific cell type without altering expression elsewhere in the body. The Cre/loxP system relies upon the ability of the phage enzyme Cre-recombinase to recognize specific DNA sequences, called loxP sites, and delete whatever is between two loxP sites (note that the genomic sequence of an entire gene is usually too big to delete, so the loxP sites need to be placed strategically in order that the gene is 'functionally' deleted). See Figure 26.1.

Thus, two genetically modified animals need to be crossed: one that has the strategically placed loxP sites 'around' the gene to be targeted; and the other that expresses Cre-recombinase in the cell type that we might wish to target. The loxP-flanked animal (or 'floxed' animal) is generated by the ES cell technique (as we are altering an endogenous gene) and the Cre-recombinase expressing gene is usually inserted using transgenic techniques (as we are adding the expression of Cre-recombinase in a specific cell type). For example, deletion of the transcription factor CREB specifically from the liver would require a 'floxed' CREB animal to be crossed with an albumin-Cre animal (albumin is a hepatocyte-specific promoter, i.e. an animal that expresses Cre-recombinase specifically in the liver).

In recent years, the Cre/loxP system has been significantly improved and variations have been introduced, making this technology even more flexible in terms of developmental time and tissue-dependent expression (the reader is referred to review articles such as Aronoff & Petersen (2006)).

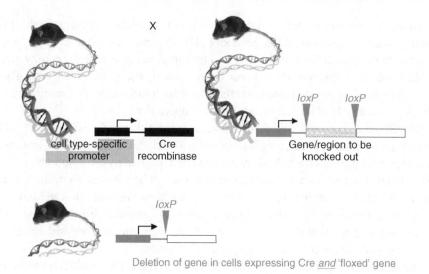

Figure 26.1 The Cre/loxP system. Mice expressing Cre-recombinase driven by a cell-type specific promoter are crossed with mice harbouring two loxP sequences either side of the genomic area to be altered. In offspring of this cross with the correct genotype, the gene of interest will be altered only in cells expressing the Cre-recombinase. Cre-recombinase recognizes the two loxP sequences and deletes whatever is between them, leaving one loxP sequence in place. Courtesy of Dr. Nina Balthasar.

26.2   Required controls, common errors and pitfalls

26.2.1   Transgenesis

As transgenesis results from random insertion of standalone pieces of DNA into the genome, it is important to demonstrate that the phenotype (physiological appearance/behaviour) of the animal is a consequence of the expression of the transgenic construct itself, and *not* of the insertion of the construct into a specific genomic locus. For example, the transgene construct may have been inserted into the coding sequence of another gene, effectively deleting this gene. Similarly, the construct might have inserted into an area of strong transcriptional expression, such that expression of the construct is influenced by surrounding genomic sequences rather than just by its own promoter.

The control for this is easy. To confirm 'insertion-independent' expression of a transgene, one needs to demonstrate that two (or more) different lines of transgenic mice (a 'line', or offspring derived from single founder animal) have identical phenotypes. As transgenesis results from random integration, the integration sites of the transgene in different founder lines will certainly be different. So, if a phenotype of the two lines is identical, it must be an effect of the transgene expression, not a consequence of its location or expression level. Thus, always check whether the authors have compared multiple lines of transgenic mice.

Transgenesis is entirely dependent on the promoter of choice, and it is often assumed that a chosen promoter contains all the control elements for appropriate time-specific (developmental time!) and region/cell type-specific expression of the gene of interest. Most of these promoters are not well defined and, while authors might anticipate that the sequences included in their transgene are sufficient to drive expression in the target cell type, this does not always happen. So, always check that the authors have carefully checked and reported their transgene expression pattern in multiple tissues and cell types – even the ones in which they do not expect expression of the transgene (see Primers 15 and 21).

We call expression of transgenes in cell types where the promoter is not normally expressed, 'exogenous' expression. Exogenous expression of the gene of interest in a tissue that has *not* been checked for expression can have a profound effect on the phenotype of the animal and can 'confound' any interpretation placed on the data.

26.2.2   Es cell techniques

Remember that ES cell techniques result in alterations to endogenous genes. If the aim is to delete a gene or insert a particular mutation in a gene, the authors must demonstrate that the gene function is, indeed, altered in the resulting animal. Therefore, a strategy for testing the function of this gene must be described, and the data should leave you in no doubt that the alteration of the genome resulted in the desired effect of altered gene function. This can be very tricky and deserves special attention.

Two other scenarios can happen, which will also have a profound effect on the interpretation of results:

1. Deletion/alteration of gene function could result in the mice not surviving. Thus, it is important to start by checking that mice of the altered gene are born at a normal Mendelian ratio, which indicates that none of the mice with the altered gene die *in utero*.

2. Deletion/alteration of gene function may result in the animals having no phenotype. This could be due to a different gene (maybe a family member) compensating for the loss of the removed gene. It is thus important to check which genes might be over-expressed as a result of the altered gene. Absence of an altered phenotype should not be taken as evidence that a gene lacks an important function – in fact, its function may be so important that there are alternatives to it. It simply means that one cannot study the function of this gene by deleting it from all tissues from day one of development.

In addition, genes that are expressed in multiple tissues might have different functions in these different tissues, i.e. effects of a gene in the liver might mask or influence the function of the same gene in the brain. It is not easy to tease these apart, so consider for yourself, have the authors thought about this and used appropriate methods to address these issues?

26.2.3   Cre/loxP

Using these techniques, two genetically modified mice are crossed to find an effect in the mice expressing both the floxed allele and the Cre-recombinase in a particular cell type or tissue. First, it is important that both original mice, 'floxed' mice and Cre-mice are 'normal' (i.e. unaffected by their genetic alteration) in every respect. The loxP sites should be placed in genomic areas where they do not interfere with the normal function of the gene. Also, Cre-recombinase expression in a certain cell type should be without effect. Have the authors demonstrated that both – the 'floxed' mice as well as the Cre mice – are by themselves comparable to wild-type mice? If these controls are missing, it is unsafe to assume that the phenotype in the double Cre-loxP animal really comes from deletion of the floxed gene in this cell type.

Has the Cre-expression pattern been checked carefully with a 'reporter' animal? In order to evaluate where Cre-recombinase is expressed throughout development, animals need to be crossed with a 'reporter' animal, which expresses eGFP when a floxed transcriptional blocker is deleted by Cre-recombinase. As this process is irreversible, expression in an adult animal eGFP indicates all cells that, during their lifespan, have at some point expressed Cre-recombinase.

Has successful deletion of the gene of interest been demonstrated satisfactorily? Placement of the two loxP sites can be quite tricky, and a decision often has to be made in terms of which exon(s) to delete to make the gene of interest non-functional. This is mostly unpredictable, and it is essential that the authors demonstrate appropriate deletion/dysfunction of the gene of interest. This will include PCR (Primer 20) and immunoblot analysis (Primer 15), but should also include a measure of function.

26.3   Complementary and/or adjunct techniques

Primer 13: Immunohistochemistry (AKA Immunocytochemistry).
Primer 15: Immunoblotting.
Primer 16: Applications of green fluorescence protein (GFP).
Primer 21: *In situ* hybridization (ISH).
Primer 22: Interference methods.

Further reading and resources

www.nobelprize.org/nobel_prizes/medicine/laureates/2007/advanced.html
Aronoff, R. & Petersen, C.C. (2006). Controlled and localized genetic manipulation in the brain. *Journal of Cellular and Molecular Medicine* **10**(2), 333–352.
Lewandoski, M. (2001). Conditional control of gene expression in the mouse. *Nature Reviews Genetics* **2**(10), 743–755.
Kos, C.H. (2004). Cre/loxP system for generating tissue-specific knockout mouse models. *Nutrition Reviews* **62**(6 Pt 1) 243–246.

27
Wireless Recording of Cardiovascular Signals

Julian FR Paton and Fiona D McBryde
Physiology & Pharmacology, University of Bristol, UK

27.1 Basic 'how-to-do' and 'why-do' section

The *in vivo* investigation of cardiovascular parameters chronically in conscious, unrestrained (stress-free) animals is a critical part of many modern day physiological studies. A huge breakthrough in the last decade has been the development of small, fully implantable transmitters which allow biological signals to be transmitted wirelessly and recorded remotely 24 hours a day. Although these techniques are initially more invasive due to the surgical implantation, there are significant advantages over older techniques such as tail cuff measurements of systolic pressure, or a tethered system whereby animals are physically connected via catheters and/or wires running inside an armoured cable to a commutator connected to the top of the cage for distribution to the recording computer. (Whitesall *et al.*, 2004; Figure 27.1).

Tail cuff allows a single point measurement of systolic pressure only. It is highly stressful for the animals, and so requires considerable habituation before reliable and representative measurements can be made (Whitesall *et al.*, 2004). Although the catheter/commutator option has the advantage of direct measurement of arterial pressure and is relatively cheap, it places additional stress on animals and can have a higher failure rate due to cable gnawing and/or wire breakages.

Signals that can be recorded using radio telemetry include arterial pressure, left ventricular pressure, ECG, EEG and sympathetic nerve activity (SNA). Heart rate can be extrapolated from an arterial pressure (inverse of inter-pulse interval) or the ECG trace (inverse of the R-R interval). The latter gives a more accurate assessment of the precise timing for each heartbeat, which optimizes spectral analysis of this signal (see below). Animals that can be instrumented for telemetric recordings

Essential Guide to Reading Biomedical Papers: Recognising and Interpreting Best Practice, First Edition.
Edited by Phil Langton.
© 2013 by John Wiley & Sons, Ltd. Published 2013 by John Wiley & Sons, Ltd.

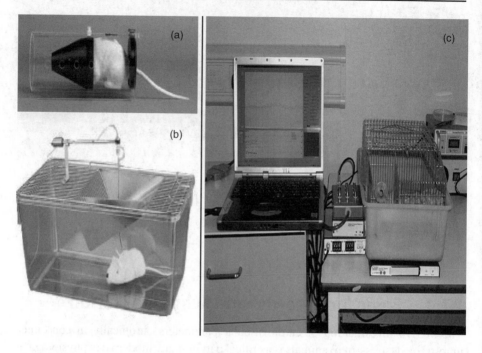

Figure 27.1 Examples of different experimental setups for assessing blood pressure. (a): Restraining tube used for tail cuff measurement of systolic pressure in the rat. (b): A rodent tethering system, with connecting wires and a commutator above the cage. (c): A telemetry setup – the instrumented rat is unrestrained in his home cage, with signals received remotely and transmitted to a recording computer. Courtesy of Professor J. Paton.

range from large (e.g. dog, pig) to small (e.g. rat, mouse). Systems are available which have a fixed battery life (from 1–4 months, depending on the size of the transmitter; see www.datasci.com) or a battery which can be recharged by induction trans-cutaneously (see www.telemetryresearch.com).

27.2   Required controls

A significant advantage of telemetry recordings is that the duration of signal recording is, in theory, limited only by the transmitter battery life. Therefore, longer duration studies can be planned and within-subject controls used. This greatly increases the statistical power of the study, allowing reductions in the number of subjects required. With *in vivo* telemetric studies, *n* numbers of six per group are typical. Usually, a time sham operated control or placebo group would be included for comparison.

With all surgeries, it is essential that at least 5–7 days elapses prior to collection of cardiovascular data post-implant of the telemetric device, to ensure a normal physiological state. This can be judged by a steady baseline (daily average) and re-establishment of a regular circadian rhythm (Figure 27.2).

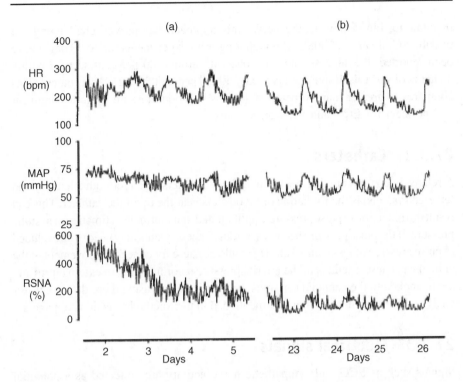

Figure 27.2   Example of data collected by telemetry. Heart rate (HR), arterial pressure (MAP) and renal sympathetic nerve activity (RSNA) in rabbits during the recovery period after implantation surgery, and three weeks later. Note that it takes at least five days for a normal circadian variation to re-establish, with strong implication on when an appropriate time point for beginning a study might be. Reproduced, with permission, from Am J Physiol Regul Integr Comp Physiol. 2001 May;280(5): R1534–45. Long-term control of renal blood flow: what is the role of the renal nerves? Barrett CJ, Navakatikyan MA, Malpas SC. © American Physiological Society.

27.3   Common problems/pitfalls in execution or interpretation

27.3.1   Surgical issues

A critical factor in successful telemetry studies is to have well-trained staff. Although many surgeries for implanting devices to detect ECG, EEG and arterial pressure are not technically difficult, training in aseptic practice and use of a surgical microscope is necessary. With blood pressure recording in rodents, the surgery is more difficult. The cannula is introduced into a large vessel such as the aorta, and held in place with tissue adhesive, as the smaller vessel sizes in rodents preclude suturing an occluding cannula in place.

One of the most difficult surgeries is to implant wire electrodes on sympathetic nerves. Because of their fragility, short lengths and small diameters, as well as inability to ensure good electrical isolation (i.e. no shunting across electrodes) while maintaining blood flow to the nerve, this approach has proved challenging but possible (Guild *et al.*, 2008). Although long-term experiments (up to 90 days) have been reported, the success rates are low, with around 80 per cent success within 48 hours of surgery, dropping to typically 50–60 per cent for a 21-day protocol. This compares to greater than 90 per cent success in blood pressure surgeries and near 100 per cent for ECG and EEG placements.

27.3.2 Catheters

A relatively common problem with fluid-filled catheters (used for both telemetry and tethered systems) is the formation of blood clots near the tip of the catheter. This can result in a reduction in pulse pressure amplitude and, thus, an underestimation of systolic pressure. If the pulse pressure becomes too small, these results may need to be excluded (a pulse pressure of less than 10 mmHg would be cause for concern). Movement of the animal can cause artefacts in the arterial pressure recording, due to external pressure being applied to the cannula; care must be taken to avoid using these data points for analysis. These points are often not mentioned in the 'methods' sections of papers.

27.3.3 Electrical signals

Signals such as ECG and sympathetic nerve activity are collected as a potential difference (voltage) between two electrodes, with reference to a ground electrode. For ECG, the electrode leads are placed subcutaneously over the chest while, for sympathetic nerve activity, the electrodes are wrapped around the nerve (e.g. renal – see Guild *et al.*, 2008) and isolated with a silicone gel. Many problems to do with recording these signals involve electrical noise. This can be due to poor electrode placement, movement of the ground lead or high levels of external electrical noise. Muscle electrical activity (EMG) is a common source of electrical noise, and experimental designs will make efforts to isolate electrodes from skeletal muscle.

For sympathetic nerve activity, a problem when investigating a change over time can be quantifying the signal output in a logical manner. The amplitude of the raw voltage signal in any given animal can differ, depending on the size of the nerves, the quality of the contact with electrodes, changes in electrode resistance and the level of 'background noise'. Several of these factors may change within an animal over time. Thus, sympathetic nerve activity must be scaled, often as a percentage change from the baseline level, or as a proportion of some other reflex change (often nasopharyngeal or baroreceptor reflex) that serves to give a presumed maximum level.

The level of background noise should be checked at the start and end of an experiment, with ganglionic blockade (a short treatment which eliminates sympathetic activity), and this background level removed from the signal. Experiments where some form of normalization has not been performed risk losing statistical power, as the variation between subjects can be high and may obscure a real effect.

27.3.4   Data volume

One issue with telemetry is that it can generate large amounts of data, and recording 24 hours a day at a high frequency will create extremely large files (e.g. half a gigabyte per rat per day at 1,000 Hz sampling rate). It is becoming increasingly common practice in the literature for 'scheduled data' recordings to be made, whereby, instead of continuous 24-hour monitoring, shorter durations of data are recorded and extrapolated to reflect the true mean. Interestingly, as little as five minutes of recording every hour can predict a 24-hour mean with less than one per cent error, so long as the recordings are staggered throughout the day (see Guild *et al.*, 2005).

27.3.5   Further analysis

From an arterial pressure or ECG recording, it is possible to extract the frequency components of these signals using a mathematical technique called 'spectral analysis'. For example, breathing causes regular fluctuations in arterial pressure (also known as Traube Hering waves), which can be detected as a frequency peak in the systolic blood pressure spectra, thus allowing respiration rate to be calculated. Similarly, by examining the various frequency components of the blood pressure or heart rate signals, inferences can be made about the balance of activity in various parts of the autonomic nervous system (Waki *et al.*, 2006).

Other algorithms which can be applied allow calculation of the baroreceptor reflex gain (sBRG, which reflects how quickly the cardiovascular system is able to oppose sudden changes in arterial pressure – caused, for example, by changes in posture or movement). All of these 'indirect' measures are useful and relatively easy to perform additions to a telemetric blood pressure protocol. However, if the primary purpose of the study was, for example, to make conclusions about sympathetic outflow, then a more direct technique (e.g. ganglionic blockade or direct nerve recordings) would provide stronger evidence. Similarly, if a particular study was investigating the baroreceptor reflex control of blood pressure, then sBRG alone would not be sufficient, as it does not allow evaluation of changes in the range or set-point of arterial pressure.

27.4   Complementary and/or adjunct techniques

Primer 34: Measurement of CO.
Primer 35: Measurement of blood flow.

Further reading, references and resources

Barrett, C.J., Ramchandra, R., Guild, S-J., Lala, A., Budgett, D.M. & Malpas, S.C. (2003). What sets the long-term level of renal sympathetic nerve activity? *Circulation Research* **92**, 1330–1336.

Guild, S.J., Barrett, C.J. & Malpas, S.C. (2005). Long-term recording of sympathetic nerve activity: the new frontier in understanding the development of hypertension? *Clinical and Experimental Pharmacology and Physiology* **32**(5–6), 433–439.

Guild, S.J., Barrett, C.J., McBryde, F.D., Van Vliet, B.N. & Malpas, S.C. (2008). Sampling of cardiovascular data; how often and how much? *American Journal of Physiology. Regulatory, Integrative and Comparative Physiology* **295**(2), R510–R515.

Waki, H., Katahira, K., Polson, J.W., Kasparov, S., Murphy, D. & Paton, J.F. (2006). Automation of analysis of cardiovascular autonomic function from chronic measurements of arterial pressure in conscious rats. *Experimental Physiology* **91**(1), 201–213.

Whitesall, S.E., Hoff, J.B., Vollmer, A.P. & D'Alecy, L.G. (2004). Comparison of simultaneous measurement of mouse systolic arterial blood pressure by radiotelemetry and tail-cuff methods. *American Journal of Physiology. Heart and Circulatory Physiology* **286**(6), H2408–H2415.

28
Electrical Stimulation Methods

Jon Wakerley

Centre for Comparative and Clinical Anatomy, University of Bristol, UK

28.1 Basic 'how-to-do' and 'why-do' section

Electrical stimulation is undertaken whenever it becomes necessary to excite a group of cell bodies or axons within the brain or spinal cord, or to excite fibres in a peripheral nerve. Electrical stimulation can be used to establish the function of particular structure within the brain, in which case a period of stimulation is applied to the region of interest while appropriate experimental measurements are made to detect a behavioural or physiological response.

Electrical stimulation can also be used to map the connections of the brain or spinal cord and to investigate how one structure affects another. In this case, the stimulation is applied while electrical recordings are taken from an anatomically and/or functionally related structure to determine how the stimulation alters neuronal firing. The evoked response can involve antidromic or orthodromic conduction (see Primer 30). Applied to peripheral nerves, electrical stimulation may be used to evoke muscle contractions or autonomic responses distally, as well as central antidromic or orthodromic responses. An example of how electrical stimulation can be used to investigate the regulation of excitatory transmission through a defined pathway within the brain is shown in Figure 28.1 .

Over the last decade, there has been a considerable expansion in the use of electrical stimulation for medical purposes. Chronic stimulation of the brain is used, for example, to treat the symptoms of Parkinsonism, while stimulation of peripheral nerves is a recognized method for alleviating pain in cancer patients. Future clinical applications of electrical stimulation may include the treatment of epilepsy, psychopathological disorders and stroke.

Essential Guide to Reading Biomedical Papers: Recognising and Interpreting Best Practice, First Edition. Edited by Phil Langton.
© 2013 by John Wiley & Sons, Ltd. Published 2013 by John Wiley & Sons, Ltd.

(a)

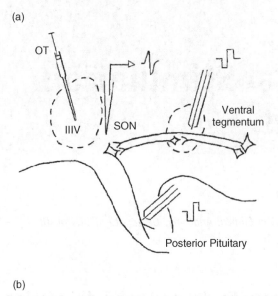

(b)

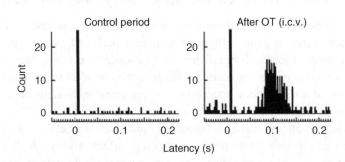

Figure 28.1 Example illustrating how electrical stimulation can be used to investigate the regulation of excitatory transmission within the rat brain in this case between the ventral tegmentum and oxytocin neurones of the supraoptic nucleus (SON) in the hypothalamus (see Cosgrave *et al.* (2000) for further details). Part (a) shows the experimental set-up. A concentric stimulating electrode is located in the neural stalk for the antidromic identification of the oxytocin neurones (see Primer 30 for a description of this technique), and a second electrode is positioned in the ventral tegmentum. A microsyringe containing oxytocin (OT) is positioned with its needle tip in the third ventricle (IIIV). Following identification of an oxytocin neurone, single pulses are applied to the ventral tegmentum every three seconds, and the response of the neurone is monitored by generating a peri-stimulus time-interval histogram (see Primer 31). Part (b) shows a typical result. Prior to giving oxytocin (left), the neurone is unaffected by the stimulation but, following oxytocin, a clear excitation is observed, indicating an autoregulatory role for this peptide in ascending transmission from the ventral tegmentum to the oxytocin neurones. Courtesy of Professor Jon B. Wakerley.

Electrical stimulation is also used in medical prostheses, particularly cochlear implants. This expansion in the medical use of electrical stimulation has lead to a resurgence of research into the theoretical aspects of this technique, in order to improve its safety and efficiency.

The two major considerations in undertaking (or, indeed, evaluating) experiments involving electrical stimulation are:

1. the design of the stimulating electrode;

2. the stimulus parameters.

28.1.1   Electrode design

By far the most common approach to electrode design is to employ a bipolar stimulating electrode, in which the anode and cathode are juxtaposed to limit the spread of current. While a monopolar electrode may sometimes be employed, with the other end of the circuit being connected to a distant part of the body or isolated tissue, this will result in a greater spread of current and more diffuse stimulation.

Most bipolar electrodes used for experimental stimulation of the brain are of a concentric design, comprising an insulated tungsten or platinum wire threaded down a steel tube, insulated on the outside except near its tip. Current is passed between the un-insulated end of the wire protruding from the tube, and the un-insulated terminal part of the tube. The shape of the resulting zone of stimulation will vary considerably, depending upon the shape of the tip (Figure 28.2).

An alternative electrode configuration, often used for clinical purposes, consisting of a series of exposed cylindrical plates (usually platinum) arranged along the tip of a semi-rigid shaft which is formed from the connecting leads of the plates. While the size and shape of the activated zone will be primary considerations, the design of electrodes for brain stimulation must also take into account the relative size of the brain, the depth of the target and (in chronic experiments) how the electrode is to be fixed in place.

Electrodes for stimulation of peripheral nerves tend to comprise a flexible sheath for fitting around the nerve, with two exposed plates arranged along its interior. Where it is necessary to deliver stimulation on a much smaller scale (e.g. to a small area of a brain slice *in vitro*), the stimulus pulses can be applied through glass micropipettes filled with sodium chloride solution, or via a pair of fine insulated platinum wires.

28.1.2   Stimulus parameters

The majority of studies employ square wave pulses, usually of duration ranging from 0.1–4 msec. Sine wave stimulation is also occasionally employed, but never direct current. Besides causing damage, the latter will result in protracted depolarization of neuronal membranes, whereby the sodium channels become inactivated and action potentials are prevented (i.e. accommodation occurs).

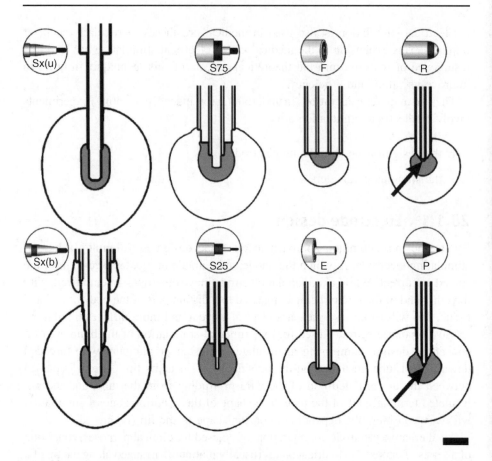

Figure 28.2 Importance of electrode shape in defining the zone of stimulation. The figure shows theoretical transmembrane potentials evoked by a 500 μA stimulation passed through concentric electrodes of various tip shapes. Characters within the circles denote the manufacturer's electrode type (see Gimsaa *et al* 2006 for details). The white and grey-filled areas represent the 1 and 10 mV limits for the induced transmembrane potential, respectively. For electrodes R and P, an additional line (arrowed) that delimits 1V is also shown. Scale bar = 200 μm. Reproduced, with permission, from Gimsaa U, Schreiber U, Habel B, Flehr J, van Rienen U, Gimsab, J (2006). Matching geometry and stimulation parameters of electrodes for deep brain stimulation experiments - numerical considerations. Journal of Neuroscience Methods 150, 212–227 © Elsevier.

As a general rule, short pulses of 1 ms or less are preferable, since they minimize current spread. Increasing the length of the pulse results in a lower activation threshold (i.e. the stimulus voltage can be reduced), but the overall amount of charge that is passed into the tissue increases as the pulse is lengthened, so there is more risk of tissue damage. Whereas monophasic pulses are the most effective, most studies involving repetitive stimulation will employ biphasic pulses (i.e. the current is reversed halfway through the pulse), because this results in zero net current flow, again reducing the risk of tissue damage. The relative efficiency and 'safety' of different pulse configurations are shown in Figure 28.3.

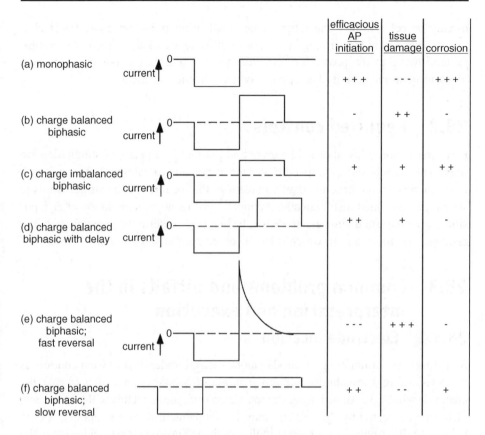

	efficacious AP initiation	tissue damage	corrosion
(a) monophasic	+ + +	– – –	+ + +
(b) charge balanced biphasic	–	+ +	–
(c) charge imbalanced biphasic	+	+	+ +
(d) charge balanced biphasic with delay	+ +	+	–
(e) charge balanced biphasic; fast reversal	– – –	+ + +	–
(f) charge balanced biphasic; slow reversal	+ +	– –	+

Figure 28.3   The relative merits of different pulse configurations in terms of their efficacy (i.e. ability to generate action potentials) versus likelihood of causing tissue damage. '+++' = best (most efficacious, least damaging to tissue or the electrode) and '– – –' = worst. The likelihood of electrode corrosion is also indicated. '+++' = most corrosion. Reproduced, with permission, from Merrill DR, Bikson M, Jeffery JGR (2005). Electrical stimulation of excitable tissue: design of efficacious and safe protocols. Journal of Neuroscience Methods 141, 171–198 © Elsevier.

The stimulation intensity (i.e. voltage) used is, to some, extent empirical – the aim being to use a minimum voltage compatible with obtaining the desired neuronal activation. Because of variations in the impedance of the electrode and surrounding tissue, it is important to monitor current flow as well as stimulus voltage. In order to maintain a consistent level of stimulation over time, it is best to employ a 'constant current' stimulator to compensate for these impedance variations.

Finally, the experimenter must decide upon the overall duration of the stimulus train and its pulse frequency. Where stimulation is applied for the study of connections (i.e. the aim is to evoke an antidromic or orthodromic response in a recorded neurone), the pulses are usually applied singly or as a short train of two or three pulses. However, if the experimenter wishes to evoke a behavioural or physiological response, then a much longer pulse train will usually be required, perhaps lasting several minutes or even hours. Pulse frequency within the train will usually be set to reflect the range of neuronal firing frequency (say, 10–50 Hz). Frequencies in the higher range (say, above 200 Hz) are likely to result in impulse failure, owing to the post-spike refractory period, and may result in a complete block to neuronal activity because of sodium channel inactivation.

28.2   Required controls

Characterization of the threshold current and optimal parameters of stimulation are important in establishing the specificity of the response and enabling comparison with other studies. In the case of brain stimulation, the location of the stimulation site should be confirmed histologically. Application of identical stimulation after repositioning of the stimulating electrode will be helpful in determining specificity of the response and the extent to which it has involved current spread to other structures.

28.3   Common problems and pitfalls in the interpretation and execution

28.3.1   Electrode location

Sometimes, the stimulating electrode can be located under direct visual control (as when recording from a brain slice *in vitro*), but for whole-brain studies *in vivo*, the electrode will be positioned using stereotaxic co-ordinates, and this will introduce an element of uncertainty. Stereotaxis may be supplemented by radiography or (in humans) MRI. Application of test stimuli to activate known pathways adjacent to the intended target may also be helpful where this evokes a readily detectable response, such as a muscle contraction. Electrode location is obviously crucial in the interpretation of experimental results, so it is essential to confirm the site of stimulation using histological methods. However, this only provides information about the position *after* the experiment has been completed.

28.3.2   Current spread

A major consideration in interpreting the effects of electrical stimulation is the extent of current spread. If the stimulation activates fibre tracts or neuronal populations lying some distance from the electrode, this will obviously negate any conclusions that link the stimulation site to a particular response. As a general rule, the current required to excite a neurone or axon increases with the square of its distance from electrode, so that, with moderate currents of around 1 mA, the zone of stimulation will be restricted to within 1–2 mm of the electrode tip (see Tehovnik, 1996). However, this represents significant spread, especially in studies involving small laboratory species, so that checking for the extent of current spread is important (see below).

28.3.3   Complex connectivity of the stimulation site

When applying stimuli to a given structure, it is inevitable that the excitation evoked in axons located either within or near the structure may reverberate (both ortho-dromically and antidromically) to other areas of the brain. Thus, a stimulus train applied to a particular region to investigate its function may sometimes evoke complex and widespread physiological responses that make the results hard to interpret. Similarly, in studies of connectivity, single pulse stimulation of a given structure may evoke highly complex responses at the recording site, perhaps comprising both inhibitory and excitatory components. Such problems are particu-larly evident when stimulating the reticular formation of the brain stem.

28.3.4   Impulse blockade

While the natural assumption is that electrical stimulation causes an increase in the activity of the surround cells and axons, this may not necessarily be the case. As already discussed, prolonged high frequency stimulation in excess of the normal firing rate may result in neural membranes becoming refractory and incapable of generating action potentials, so that activity is actually prevented.

28.3.5   Tissue damage

Electrical stimulation carries a risk of causing tissue damage, especially when employing currents in the higher range, and this may be accompanied by electrode corrosion. Damaging effects of the stimulation will result in a diminution or loss of the evoked response, which may be temporary or permanent. In extreme cases, tissue damage will also be evident histologically. The cause of tissue damage is complex and may include deposition of metallic ions and/or generation of hydrogen ions, free radicals and gas bubbles. Mass activation during stimulation may also cause cytotoxic effects arising from an excess of intracellular calcium or extrac-ellular glutamate. In order to minimize the risk of tissue damage, stimulation currents should be kept as low as possible, and stimulus pulses should be short and charge balanced (see foregoing discussion; Figure 28.3). Finally, it should be noted that stimulating electrodes used within the intact nervous system of laboratory species are relatively large and will cause damage as they are lowered into position.

28.3.6   Stimulus artefacts

When stimulation is combined with electrical recording, a common problem is that the stimulus artefact swamps the recording or 'saturates' the recording pre-amplifier, so that the evoked signal is lost, particularly if it has a short latency. This problem can be reduced by using very short stimulus pulses and by employing stimulus isolation units that are electrically isolated from the recording apparatus. Such isolation units are driven by a programmable timing device that can not only produce various patterns of stimulation, but can also produce synchronizing pulses for other equipment, such as oscilloscopes and computers.

28.3.7 Unspecific responses

While they may seem trivial, one should always be aware of the possibility of 'nonspecific' responses arising from things such as muscle contraction or cardiovascular changes evoked by the stimulation. This can be a particular problem in experiments involving electrophysiological recordings, where a sudden displacement of the recorded cell may cause its mechanical stimulation.

28.4   Complementary and/or adjunct techniques

As an alternative to electrical stimulation it is possible to use microinjections of glutamate or NMDA to excite a particular structure within the brain; these excitatory agents can also be topically applied to brain slices *in vitro*. This chemical stimulation has the advantage that only cell bodies are excited, since axons do not contain glutamate receptors. However, a disadvantage is that chemical excitation may be inconsistent in its intensity and have an unpredictable time course. Other related techniques include microelectrode recording and magnetic stimulation.

Primer 9: Microelectrode recording.

Primer 29: Extracellular recording.

Primer 31: Event-triggered spike averaging.

Further reading and resources

Butson, C.R. & McIntyre, C.C. (2006). Role of electrode design on the volume of tissue activated during deep brain stimulation. *Journal of Neural Engineering* **3**, 1–8.

Cosgrave, A.S., Richardson, C.M. & Wakerley, J.B. (2000). Permissive effect of centrally administered oxytocin on the excitatory response of oxytocin neurones to ventral tegmental stimulation in the suckled rat. *Journal of Neuroendocrinology* **12**, 843–52.

Gimsaa, J., Habela, B., Schreiberb, U., van Rienenb, U., Straussc, U. & Gimsac, U. (2005). Choosing electrodes for deep brain stimulation experiments – electrochemical considerations. *Journal of Neuroscience Methods* **142**, 251–265.

Gimsaa, U., Schreiber, U., Habel, B., Flehr, J., van Rienen, U. & Gimsab, J. (2006). Matching geometry and stimulation parameters of electrodes for deep brain stimulation experiments – numerical considerations. *Journal of Neuroscience Methods* **150**, 212–227.

Kuncel, A.M. & Grill, W.M. (2004). Selection of stimulus parameters for deep brain stimulation. *Clinical Neurophysiology* **115**, 2431–2441.

Merrill, D.R., Bikson, M. & Jeffery, J.G.R. (2005). Electrical stimulation of excitable tissue: design of efficacious and safe protocols. *Journal of Neuroscience Methods* **141**, 171–198.

Patterson, M.M. & Kesner, R.P. (1981). *Electrical stimulation research techniques.* Academic Press, New York.

Tehovnik, E.J. (1996). Electrical stimulation of neural tissue to evoke behavioral responses. *Journal of Neuroscience Methods* **65**, 1–17.

29
Extracellular Recording

Jon Wakerley

Centre for Comparative and Clinical Anatomy, University of Bristol, UK

29.1 Basic 'how-to-do' and 'why-do' section

Extracellular recording is used to study the electrical activity of neurones in order to investigate the input-output relations of a neuronal circuit, or how the activity of a population of neurones within a particular neuronal structure is correlated with a particular physiological response. As the term implies, extracellular recordings are undertaken with the electrode positioned outside of the neurone being studied. In fact, extracellular recordings may be derived from individual neurones (called 'single-unit' activity) or from several neurones recorded simultaneously (called 'multi-unit' activity), depending upon the tip diameter of the electrode. In both cases, the recordings consist of a series of action potentials (also called 'spikes') superimposed upon a background noise.

Whereas this primer will focus mainly on the extracellular recording of action potentials, it should be noted that the term 'extracellular recording' also encompasses the recording of 'field' potentials (also called 'evoked potentials'). Field potentials are monitored with a large low impedance electrode, with the recording system set to pick up low frequency voltage fluctuations rather than action potentials. Field potentials represent the summed activity (both action potentials and synaptic potentials) occurring simultaneously within a large population of neurones, and they are usually evoked experimentally by electrical stimulation.

Extracellular recording of action potentials from single neurones is generally an easier technique than intracellular recording (Primer 9) or whole-cell patch clamping (Primer 11), because there is much less risk that the neurone will be damaged or lost because of mechanical movement. Moreover, extracellular recording microelectrodes are easier to manufacture than intracellular microelectrodes, and the technique is more tolerant of variations in electrode characteristics.

Essential Guide to Reading Biomedical Papers: Recognising and Interpreting Best Practice, First Edition.
Edited by Phil Langton.
© 2013 by John Wiley & Sons, Ltd. Published 2013 by John Wiley & Sons, Ltd.

However, it should be remembered that extracellular recording only provides information about voltage changes associated with supra-threshold membrane events (i.e. action potentials), so any conclusions concerning the underlying membrane conductance changes or synaptic currents have to be inferred indirectly. Nowadays, most electrophysiological studies on isolated tissues use patch clamping, but extracellular neuronal recording is still employed commonly for *in vivo* studies where tissue instability (due to blood pressure pulsations) makes it difficult to use intracellular or patch clamping recording.

Owing to the advent of sophisticated methods for the recording and discrimination of multiple neurones simultaneously, and of techniques for undertaking recordings in awake and freely moving animals, extracellular recording remains a powerful approach for studying the integrative functions of the nervous system. The opportunity to monitor multiple neurones simultaneously during behaviour or cognition is particularly useful, especially when combined with advanced paradigms for cross-correlation and pattern analysis.

Microelectrodes for extracellular recording of individual neurones are typically composed of glass micropipettes filled with 0.5 M sodium acetate. For single cell recording, the tip of the electrode must be sufficiently fine (2–3 μm), with an impedance of at least 2–3 MΩ, otherwise the electrode will pick up signals from more than one cell simultaneously. As an alternative to glass, electrodes can be made by etching a fine metal wire (usual steel or tungsten) that is then insulated up to its tip using epoxy resin.

Once manufactured, the electrode needs to be fixed in a micromanipulator so that it can be stereotaxically positioned to track through the area of interest. The electrode can be lowered towards its target quite rapidly but, once the target is reached, the electrode must be advanced very slowly (i.e. in steps of a few microns) until an individual recording is detected, indicated by the appearance of action potentials on the oscilloscope trace. For obtaining a good long-term recording, the electrode must be located sufficiently close to the neurone to pick up the signal, but not too close as to cause mechanical damage. With practice, it is surprisingly easy to obtain stable recordings from individual neurones, and such recordings can often be held for several hours.

Extracellular electrodes designed for single unit recording only detect action potentials from the cell body and rarely pick up axonal spikes. However, the possibility of obtaining recordings from axons, rather than cell bodies, is increased when using large-tipped, low impedance electrodes.

Extracellular action potentials are usually of less than 1 mV in amplitude so, for extracellular recording, it is necessary to use a high-gain ($\times 1000$) preamplifier. Extracellularly recorded action potentials usually have a triphasic (positive, negative, positive) waveform, but this varies according to a number of factors, such as: the position of the electrode in relation to the cell (see Figure 29.1); the capacitance characteristics of the electrode; and the pass-band characteristics of the recording system.

Unless the recording conditions are very carefully defined, one must be cautious in drawing too many conclusions from action potential waveform when undertaking

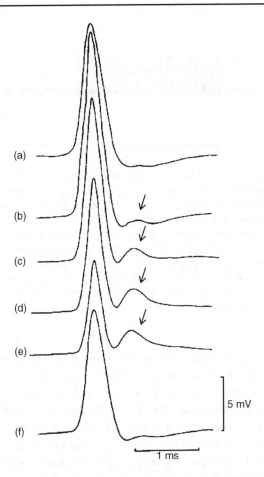

Figure 29.1 Extracellular action potentials recorded from a hypothalamic magnocellular neurone to illustrate how the waveform of the spike is altered as the electrode is moved towards the cell. The top trace (a) shows the waveform when the neurone was first encountered; traces (b–e) show how the waveform changed as the recording electrode was advanced in 20 μm steps. The final trace (f) shows the waveform after the electrode had been withdrawn to its original position. The second positive wave on traces (b–e) (arrowed) represents the calcium spike of the dendrite which rapidly follows the larger sodium spike of the soma. Each sweep shown is the average of approximately 50 spikes. From Mason WT, Leng, G (1984). Complex action potential waveform recorded from supraoptic and paraventricular neurones of the rat: evidence for sodium and calcium spike components at different membrane sites. Experimental Brain Research 56, 135–43. © Springer Publishing.

extracellular recording. That said, the action potentials of some neurones have particular features that can help in their identification – for example, 5HT neurones of the dorsal raphe display action potentials that are of an unusually long duration. Hypothalamic magnocellular neurones display two positive components of their action potentials, one being derived from the sodium spike of the soma, the other from the calcium spike of the dendrites (see Figure 29.1).

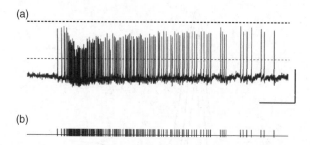

Figure 29.2 Use of a window discriminator to convert an extracellular recording of spike activity into a series of pulses that can be fed into a computer for further analysis. (a) shows a train of extracellularly recorded action potentials from a hypothalamic oxytocin neurone during a burst of high frequency firing that is characteristic of these neurones (scale is 0.5 s and 0.75 mV). The dotted lines indicate the upper and lower settings of the window discriminator used to analyze the trace; any deflection of the trace that has a peak falling between these two levels triggers a short pulse. (b) shows the series of pulses derived from the discriminator. Note how the spike amplitude almost falls below the window during very rapid firing, emphasizing the importance of the window settings in obtaining an accurate analysis of the activity. From Dyball REJ, Leng G (1986). Regulation of the milk ejection reflex in the rat. *Journal of Physiology* **380**, 239–256. © Physiological Society.

Most researchers using extracellular recording will pay little attention to the waveform of the recorded action potentials. Instead, the focus will be on the frequency and/or patterning of the spikes. For such analysis, the recorded action potentials must be discriminated from the background noise and converted into a series of pulses (each representing a spike), either using a window discriminator that detects signals between two preset levels (see Figure 29.2), or by appropriate computer software that recognizes the action potentials from a preset template (see Primer 31 for more details).

The latter approach relies upon sophisticated algorithms that allow several neurones to be discriminated from a multiunit recording, a process referred to as 'spike sorting' (see Figure 29.3).

Pulses representing extracellular spikes are then subjected to computer analysis to generate, for example, a sequential histogram of second-by-second firing rates, an autocorrelogram or a peri-stimulus time-interval histogram (see Primer 31). Many articles based on extracellular recording will just display these derived data, without representation of the original action potentials.

In addition to glass micropipettes or etched wire, a number of other extracellular electrodes have been developed for monitoring single- and multi-unit activity in the central or peripheral nervous system. These are briefly described below.

29.1.1 Microwire electrodes

These have been developed mainly for use in chronic preparations. Typically they comprise 6–8 fine (around 25 μm) insulated platinum-iridium wires, each with an exposed tip that will pick up multiunit activity rather than single units. To enable

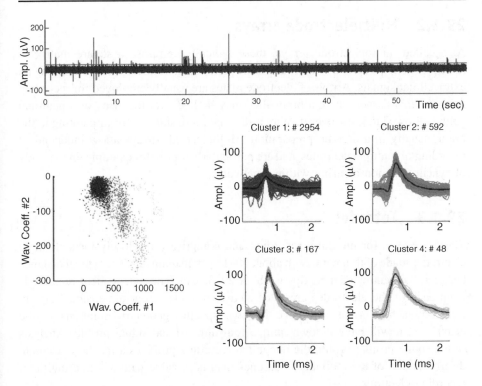

Figure 29.3   Example to illustrate how spike sorting can be used to discriminate individual neurones within a multi-unit trace. The top trace shows an extracellular multi-unit recording, in this particular case obtained with a microwire electrode positioned in the hippocampus. The red line above the trace indicates the threshold used for signal detection (i.e. all signals below this line are ignored). Waveforms above the line were analyzed to generate two 'wave coefficients' for each spike, each coefficient being based on particular aspects of the shape of the action potential. The lower graph on the extreme left shows a plot whereby each action potential (represented by a dot) is positioned according to its two coefficients. Note how this results in four distinct clusters of spikes (indicated by the different colours), which represent four distinct waveforms that have been discriminated. The averaged waveforms (in black) and superimposed individual waveforms (coloured) for these four spike clusters are shown in the lower right of the figure. Numbers above the waveforms refer to the number of spikes that have been included. Reproduced, with permission, from Quian Quiroga R, Reddy L, Koch C, Fried I (2007). Decoding visual inputs from multiple neurons in the human temporal lobe. Journal of Neurophysiology 98, 1997–2007. © American Physiological Society. *A full colour version of this figure appears in the colour plate section.*

implantation, the wires are held within a rigid tube or cemented together into a semi-rigid bundle, using an adhesive which dissolves in the brain to allow dispersion of the wires. The wires are attached to an anchoring and connection system on the skull, which may include a means of adjusting their depth. Where single channel recording is employed, each wire is sampled in turn to find the strongest signals. Alternatively, multichannel recording may be used, whereby signals are recorded independently and simultaneously from several of the wires.

29.1.2 Multielectrode arrays

Also called microelectrode arrays, these typically comprise a square matrix of multiple (perhaps up to 64) microelectrode plates or short microwires for picking up extracellular signals. Arrays of electrode plates are usually employed for recording from *in vitro* tissue, where a brain slice may be laid over the array or a neuronal culture allowed to grow over it. Wire arrays are more suitable for implanting in the brain, usually in a chronic preparation. Multielectrode arrays allow independent recording of multiple neurones, and are particularly useful, for example, in the study of synchronization of firing within a network.

29.1.3 Tetrodes

These are used for multiunit recording and comprise four closely-located single electrodes made of thin wire, each about 10–15 μm in diameter. The typical distance between the centres of their tips is about 20–30 μm. The independent recordings from each of the electrodes are integrated to provide a four-dimensional signal. Because each neurone will have its own specific geometric relations to the electrodes, it will have its own unique four-dimensional signal profile. Analysis of the signal profiles within the integrated recording provides a far more accurate discrimination of the individual neurones than is possible with a single multiunit recording electrode.

29.1.4 Carbon fibre electrodes

At their simplest, carbon fibre electrodes comprise a fine (around 7 μm) carbon fibre located within a glass micropipette. Carbon fibre electrodes have lower impedance for their tip size, compared to conventional glass or metal electrodes, and this improves signal amplitude and reduces noise. It is also possible to manufacture more sophisticated multichannel carbon electrodes comprising multiple fibres, each enabling an independent recording. Such electrodes can be used in the same way as tetrodes to record multiunit activity in several dimensions, allowing easier separation of the individual neurones.

29.1.5 Cuff and suction electrodes

These are used for recording from peripheral nerves. Cuff electrodes enclose a short length of the nerve in a cylindrical sleeve which has a number of small electrode contacts fixed to its interior wall. The exclusion of extraneous signals (e.g. from adjacent muscles) is achieved by using differential recording, with the earthing contact also located within the sleeve. Suction electrodes are designed to hold on to the ends or sides of nerves using suction. They comprise a glass micropipette with a diameter that will accommodate the whole cut nerve in its tip, or will allow the tip to be applied to the unsevered nerve. Suction is achieved through a three-way stopcock connected to a syringe, and electrical contact with the nerve is achieved via the fluid medium (balanced salt solution) within the pipette.

29.2   Required controls

29.2.1   Location of recording site

In order to be able to interpret the recorded neuronal activity, it is essential to know the location of the recordings. When using glass electrodes, this can be achieved by filling the micropipette with a dye which is then iontophoretically ejected into the surrounding tissue at the end of the recording. In the case of metal electrodes, the recording position can be marked by passing current through the electrode to make a small lesion, or (if the electrode is made of iron) to leave a metallic deposit that can be detected histochemically.

29.2.2   Neuronal identification

The usefulness of extracellular recording is hugely increased if one knows the functional identity of the neurone from which the activity was derived. In contrast to intracellular recording, it is usually not possible to ascribe extracellularly recorded activity to a single histologically identified neurone.

There is a method called 'juxtacellular labelling' which enables a cell to be filled with dye using an extracellular electrode (Allers & Sharp, 2003), but this is by no means a routine technique. Instead, extracellular studies must rely upon physiological methods to identify (or at least to categorize) the recorded neurones. This will be based on antidromic identification (see Primer 30) to determine their axonal projection, perhaps coupled with a series of physiological and/or pharmacological tests, and analysis of its firing pattern to build up putative identity profiles.

29.3   Common problems and pitfalls in the interpretation and execution

29.3.1   Recording artefacts

Because of the signal amplification, extracellular records can be easily contaminated with artefacts derived from electronic noise, static and 50-cycle 'hum'. Extraneous biological signals derived from muscle activity or the ECG can also be a problem. Appropriate earthing and screening, as well as filtering of the record signal, and careful adjustment of the window discriminator, can all help prevent artefacts from spoiling the recording. However, if the recorded action potentials are of small amplitude, it can be very difficult to avoid the inclusion of some artefacts in the derived data.

29.3.2   Sampling bias

The detection of a neurone during extracellular recording is done entirely blind and has been likened to 'fishing', in that the experimenter cannot predict when a neurone will appear. If a neurone is completely silent, it cannot be detected by extracellular recording, other than by evoking an antidromic spike, and this requires knowledge of its axonal projection. Small neurones, as well as neurones with high firing rates, tend to display low amplitude spikes which may be overlooked by the experimenter. Hence, the sample may be biased against some types of neurones (see Primer 2).

29.3.3   Spike discrimination and sorting

Even with fine-tipped electrodes designed for single cell recording, it occasionally happens that more than one neurone will appear on the oscilloscope trace and, as described above, electrodes may be used that are deliberately designed to pick up recordings from multiple neurones. If the multiple recordings involve just two (or possibly three) neurones, with spikes of very different amplitudes, it can be possible to separate them using a window discriminator. However, it is more likely that the action potentials of the different neurones will be of quite similar amplitude, so that discrimination will also need to take into account differences in spike shape. A number of commercial packages have been developed to enable 'spike sorting' in multiple recordings, and these are based on algorithms that sample the spikes and generate a series of templates to which each spike is compared (see Figure 29.3). However, fluctuations in the amplitude or waveform of the spikes, or overlapping of synchronous spikes, may impair the accuracy of spike separation.

29.3.4   Defining a response

Although somewhat beyond the scope of this primer, an issue which often arises in studies based on extracellular recording concerns the criteria used for defining a significant neuronal response. For example, statistical tests may indicate that a change in firing from 1.2 to 1.7 spikes/second is significant at the five per cent level, but this does not necessarily mean that a change in firing of 0.5 spikes per second is *physiologically* significant. It is inevitable that some extracellular recording studies include rather arbitrary criteria for deciding that a neurone is responsive to an experimental manipulation.

29.3.5   Chronic tissue response to the recording electrode

A particular problem in undertaking long-term extracellular recordings lasting over several weeks is that the presence of the recording electrode may evoke a glial reaction which will cover the electrode with a cellular matrix and reduce signal detection. The use of biologically inert conductive films for coating chronically implanted recording electrodes is a potential solution to this problem.

29.4   Complementary and/or adjunct techniques

Complementary techniques commonly used with extracellular recording are:
Primer 28: Electrical stimulation.
Primer 30: Antidromic identification.
Primer 31: Event-triggered averaging.

Extracellular recording is often used in neuropharmacological studies, where it is necessary to apply drugs locally to the neurone. This can be undertaken using multi-barrelled glass microelectrodes, with one barrel being used for recording and the remaining barrels being used for the iontophoretic application of drugs. As an alternative approach, the glass recording electrode can be fixed to a second glass micropipette, from which a drug can be locally applied by pressure ejection.

Further reading and resources

Allers, K.A. & Sharp, T. (2003). Neurochemical and anatomical identification of fast- and slow-firing neurones in the rat dorsal raphe nucleus using juxtacellular labelling methods *in vivo. Neuroscience* **122**, 193–204.

Dyball, R.E.J. & Leng, G. (1986). Regulation of the milk ejection reflex in the rat. *Journal of Physiology* **380**, 239–256.

Gray, C.M., Maldonado, P.E., Wilson, M. & McNaughton, B.L. (1995). Tetrodes markedly improve the reliability and yield of multiple single-unit isolation from multi-unit record-ings in cat striate cortex. *Journal of Neuroscience Methods* **63**, 43–54.

Lehew, G. & Nicolelis, M.A. (2008). State-of-the-art microwire array design for chronic neural recordings in behaving animals. In: Nicolelis, M.A. (Ed.). *Methods for Neural Ensemble Recordings*. 2nd edition. CRC Press, Boca Raton, FL.

Loeb, G.E. & Peck, R.A. (1996). Cuff electrodes for chronic stimulation and recording of peripheral nerve activity. *Journal of Neuroscience Methods* **49**, 5–103.

Mason, W.T. & Leng, G. (1984). Complex action potential waveform recorded from supraoptic and paraventricular neurones of the rat: evidence for sodium and calcium spike components at different membrane sites. *Experimental Brain Research* **56**, 135–143.

Millar, J. & Pelling, C.W.A. (2001). Improved methods for construction of carbon fibre electrodes for extracellular spike recording. *Journal of Neuroscience Methods* **110**, 1–8.

Nicolelis, M.A. & Ribeiro, S. (2002). Multielectrode recordings: the next steps. *Current Opinion in Neurobiology* **12**, 602–606.

Piironen, A., Weckström, M. & Vähäsöyrinki, M. (2011). Ultrasmall and customizable multichannel electrodes for extracellular recordings. *Journal of Neurophysiology* **105**, 1416–1421.

Quian Quiroga, R., Reddy, L., Koch, C. & Fried, I. (2007). Decoding visual inputs from multiple neurons in the human temporal lobe. *Journal of Neurophysiology* **98**, 1997–2007.

30
Antidromic Identification

Jon Wakerley

Centre for Comparative and Clinical Anatomy, University of Bristol, UK

Functional studies of the nervous system frequently involve electrical recordings taken from individual neurones to study changes in their firing during a particular physiological or behavioural response. However, interpreting the functional significance of a change in the electrical activity of a recorded neurone is virtually impossible unless one has some idea of its axonal projection, be it to another part of the central nervous system or to a peripheral target, such as a skeletal muscle.

30.1 Basic 'how-to-do' and 'why-do' section

A neurone's axonal projections can be demonstrated using intracellular labelling, combined with retrograde tracing. The information is, however, only available after completion of histological analysis, usually several days after the recording experiment. The major advantage of the antidromic identification technique is that it provides a simple means of identifying the axonal target of a neurone *at the time that it is being recorded*. It follows that antidromic identification enables the experimenter to select a particular population of neurones on the basis of their efferent projection pathway or axonal target. Thus, besides helping with data interpretation, another benefit of using antidromic identification is that it enables the experimenter to avoid analyzing neurones that have no relevance to the study, so improving the efficiency of data collection.

Antidromic identification can be undertaken in conjunction with both extracellular and intracellular recording and can be used in isolated tissues, such as brain slices, as well as in whole animal preparations. Antidromic identification relies upon the fact that action potentials can travel in any direction on the neuronal membrane. Hence, if the axon of the neurone is stimulated, usually near its terminal, the action potential that is initiated will propagate in the reverse or '*wrong*' direction (i.e.

---

[1] **Soma** = cell body.

Essential Guide to Reading Biomedical Papers: Recognising and Interpreting Best Practice, First Edition. Edited by Phil Langton.
© 2013 by John Wiley & Sons, Ltd. Published 2013 by John Wiley & Sons, Ltd.

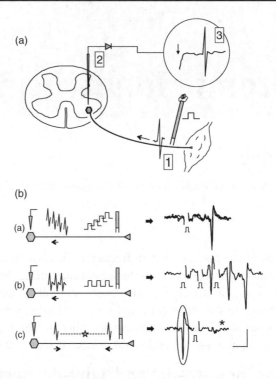

Figure 30.1 Diagram illustrating the principles and the data obtained from antidromic identification experiments. Part (a) shows the stages involved in the antidromic identification of a neurone (in this case a motoneurone in the ventral horn of spinal cord). (1) A stimulus pulse is applied to the peripheral nerve carrying the neurone's axon in order to evoke an action potential; this then passes antidromically up the axon to invade the cell body. (2) The antidromic action potential is recorded (in this case extracellularly) by a microelectrode positioned adjacent to the cell body. (3) A storage oscilloscope is used to display the antidromic response, which occurs at a fixed time delay following the stimulus artefact (arrowed). Part (b) shows the three standard tests to confirm the validity of the response, including typical oscilloscope traces (on the right of the figure). (a) Constant latency is tested by evoking 4–5 successive antidromic spikes, which are superimposed to show that there is no 'jitter' in the response. (b) Frequency following is tested by applying a train of three or more stimulus pulses and observing a fixed latency spike for each pulse. (c) Collision is tested by applying a stimulus pulse immediately after a synaptically mediated (i.e. othrodromic) spike. The orthodromic and antidromic spikes then collide on the axon, so that no antidromic response is recorded at the soma. In the example illustrated, the pulse was applied approximately 4 ms after the orthodromic spike (circled); the expected time of the antidromic spike is indicated by the asterisk. (Scales: 5 msec and 50 microvolts). Courtesy of Professor Jon B. Wakerley.

antidromically) to invade the [1]soma. The principles of the technique (see Lipski (1981) for review) are illustrated in Figure 30.1a.

It is often helpful to have some insight into the practical aspects of making an experimental observation if one's aim is to judge the quality of execution in a journal article. Imagine one is making recordings from the ventral horn of the spinal cord and one wants to be able to check whether the recorded neurone is actually a motoneurone projecting out to a peripheral nerve (say the sciatic nerve). To achieve this using antidromic identification requires dissecting out the sciatic nerve and placing it over a bipolar stimulating electrode. After a recording electrode has been positioned in the ventral horn to record an individual neurone, single shocks are delivered at low frequency (say one per second) to the sciatic nerve, while the spike activity of the neurone is recorded. If the neurone being recorded sends its axon through the sciatic nerve, it will display an antidromically evoked spike with a consistent latency after each stimulus. The precise latency will obviously be the conduction velocity of the axon and the distance to be travelled. Visualization of the antidromic spike is achieved using a digital oscilloscope or computer equipped with an analogue-to-digital (A-D) interface. In either case, the trace would be triggered to coincide with the application of the stimulus, with the expectation that the spike should occur at a particular point on the recorded trace. It is important to remember that the antidromic spike will occur in addition to any synaptically mediated ongoing spike activity of the neurone.

Usually, when recording, one may first pick up normal ongoing activity, then test the antidromic response. However, if the neurone is silent (i.e. no 'orthodromic' activity), then the antidromic spike will be the only means of detecting an extracellularly recorded neurone. Hence, another important advantage of anti-dromic identification, at least when using extracellular recording, is that it allows the detection of silent neurones.

The antidromic identification technique can be used wherever there is a defined efferent pathway, provided the pathway is anatomically suited to the application of stimulus pulses at sufficient distance from the recording site (i.e. the latency of antidromic response is long enough to avoid obliteration by any electrical artefact created by the stimulating electrodes). There are many other situations where the antidromic identification technique might be used, such as in the identification of hypothalamic neurones projecting to the posterior pituitary, or hippocampal pyr-amidal neurones projecting through the fornix to the mammillary bodies.

30.2 Required controls

When using the technique of antidromic identification, one must always keep in mind the need to demonstrate that the evoked response is truly the result of antidromic conduction, and does not arise by other means – most notably through stimulation of an afferent pathway to the neurone (i.e. that it is *not* an orthodromic response). By convention, three tests are usually applied to confirm that a neurone is being antidromically activated from the stimulus site (see Figure 30.1b).

30.2.1 The constant latency test

The first of these tests is the '*constant latency*' test. Superimposing antidromically evoked action potentials using a digital oscilloscope checks variability in the latency of the antidromic response. Synaptically evoked responses will display '*jitter*', whereby the latency varies by about 0.5 ms, whereas a true antidromic response will typically display a latency that is absolutely constant. One rare exception is if the axon branches close to site of stimulation, such that each stimulation pulse may fire either the shorter or longer branch. This will cause the response to '*jump*' between two clear latencies, but there will be no jitter.

30.2.2   The frequency following test

Because synaptic responses involve conversion of an electrical into a chemical signal that moves across the synaptic cleft, they do not follow high frequency stimulus trains reliably. Thus, when two pulses are given with an interpulse interval of, say, 10 msec (100 Hz), the second spike is often lost or will be variable in its latency. Antidromic responses, on the other hand, should *reliably* follow short trains at a frequency of 100 Hz.

30.2.3   The collision test

The third, and most important, test is the '*collision*' test. The basis of this test is to demonstrate that the antidromically-evoked spike can be made to 'collide' with a natural orthodromic spike coming down the axon. To undertake this test, the stimulator is set up in such a way that a spontaneous spike on the neurone immediately (within 0.1 msec) triggers the stimulator to deliver a pulse to the stimulation site. When this happens, there will be one action potential moving orthodromically down the axon and another moving antidromically up the axon. These two '*collide head on*' and both are extinguished by the inactivation of voltage-gated sodium channels in the wake of each action potential. The result is, therefore, that no antidromic spike is recorded at the soma.

The importance of the collision test is that it shows that spontaneous orthodromic and stimuli-evoked antidromic spikes are occurring on the same neurone. This does not necessarily have to be the case. For example, one might be recording simultaneously from two adjacent cells, one of which is silent but showing antidromic spikes, while the other is spontaneously active but not responding to the stimulus. In this case, the collision test will not work. A further confirmation of a positive collision test is achieved by calculation of the '*collision interval*', namely the latency of the antidromic response plus the refractory period of the axon at the stimulation site. Application of the triggered pulse at any time during this interval following a spontaneous action potential should result in collision but, if the pulse is delayed beyond the collision interval, then the collision test will fail.

Some studies have involved antidromic identification from more than one site in order to detect neurons with branching axons and dual projection targets (Inyushkin *et al.*, 2009). Here, the collision test can be modified to enable an estimate the distance of the branch point from the cell body. Instead of looking for collision between an orthodromic and an antidromic spike, the experimental equipment is set up to detect collision between an antidromic spike initiated on one axon with an antidromic spike initiated on the other axon. Collision will only occur if the stimulus pulses to each axon are applied within a fixed time window; the difference between the duration of this window and the summed latencies from the two stimulation sites provides an indication of the position of the axonal branch point.

30.3   Common problems and pitfalls in the interpretation and execution

Antidromic identification is a relatively straightforward technique, but it is important to undertake the tests mentioned above to confirm that the identified neurone is indeed displaying an antidromic response, and is not being stimulated by other means. Additionally, the antidromic identification technique is only valid if the stimulating electrode is correctly positioned within the axonal target site of interest.

In some cases, the positioning of the stimulating electrode can be confirmed by observing a physiological response following the application of a train of pulses, an example being the recording of an intramammary pressure response following stimuli applied to the neurohypophysis (Lincoln & Wakerley, 1974). A further complication to keep in mind is that the axon of the antidromically identified neurone may not actually terminate within the stimulated structure, but may pass through it en route to another part of the brain. Swadlow (1998) has provided a detailed review of the various problems which may arise at each stage in the process of antidromic identification, and these are summarized below.

30.3.1   Detection and recording of the candidate neurone

Recording electrodes having a particular tip size or impedance may favour the detection of some types of neurones over others, depending on their size or microanatomy. Thus, some neurones may not be detected, even though they are displaying antidromic spikes. Neurones with very low levels of spontaneous firing can also be hard to detect by extracellular recording and, for this reason, stimulus shocks are usually applied every 2–3 seconds as the recording electrode is advanced. This ensures that antidromically responding neurones are not overlooked, despite the absence of orthodromic spikes. Some neurones can be excited easily by afferent stimulation (e.g. by stimulation of the receptive field in the case of sensory neurones), and this can be exploited to increase their firing, making them easier to detect.

30.3.2   Activation of the efferent axon

The challenge here is to activate all of the efferent axons in the target of interest, without using currents of such magnitude that adjacent structures are stimulated, or the axons suffer electrolytic damage. Since small axons have a higher threshold than larger axons, the use of sub-optimal currents may result in the unintentional selection of particular subtypes of neurones. The duration of stimulus pulses is usually in the order of 0.5–1.0 msec, and biphasic (positive-negative) pulses are often used to minimize electrolysis at the stimulation site. Shorter pulses reduce the stimulus artefact, but they may not adequately excite all the axons without using higher stimulus intensities, so there is a compromise to be found. The choice of electrode configuration will depend upon the nature of the structure being stimulated but, in most cases, a bipolar concentric electrode is optimal.

30.3.3   Conduction back to the soma

Factors that can prevent or impair antidromic conduction include:

1. The existence of axonal branches, especially where the distal branches are finer than the main axonal trunk, making failure more likely;

2. Inability of the antidromic axonal spike to activate a somadendritic spike, owing to the somadendritic region being hyperpolarized;

3. Random collision with orthodromic spikes, especially in the case of neurones with a high level of orthodromic activity. In the latter case, the application of double pulses will increase the likelihood that at least one spike reaches the soma.

30.3.4   Visualization of the antidromic spike

Antidromic activation of multiple neurones may result in a multiunit response and/or field potential at the site of recording, particularly with low impedance electrodes. While this can be helpful in confirming that the recording electrode has reached its correct location, there is a risk that these multi-neurone potentials may obscure the individual antidromic spikes related to the neurone of interest. A prolonged stimulus artefact can also obscure the antidromic spike, particularly if the response has a short latency.

Such problems can be minimized by using recording electrodes with a sufficiently fine tip and high impedance to prevent the detection of signals from multiple neurones and, similarly, adjusting the [2]band-pass characteristics of the recording electrode can help to minimize the stimulus artefact. However, one must keep in mind that a recording system optimized to suppress stimulus artefacts and evoked potentials may be less than ideal for obtaining stable recordings from single neurones.

---

[2] **'Band-pass'** characteristics describe the filtering properties that can minimize either high- or low-frequency events, depending on the desired outcome.

30.3.5   Discrimination of antidromic versus orthodromic responses

As already mentioned, some neurones have such a low level of spontaneous firing that this precludes a collision test to validate their antidromic response. In such cases, one may have to rely upon the demonstration of constant latency, but slowly conducting neurones with fine axons may show a degree of latency variability arising from the residual effects of recently conducted orthodromic spikes. Furthermore, fine non-myelinated axons may be unable to conduct trains of antidromic spikes reliably, owing to their membrane becoming refractory. Such problems can result in valid antidromic responses being discounted by the inclusion criteria that has been adopted, especially in the absence of a collision test.

30.4   Complementary and/or adjunct techniques

Antidromic identification involves the use of both electrical recording and stimulation techniques, and these are reviewed elsewhere in this book. While antidromic identification can, in theory, be used to map previously unknown connections within the brain, it is far more efficient to first investigate efferent connections by conventional neuroanatomical tracing techniques, and then use antidromic identification for functional studies. Tracing techniques usually rely upon retrograde transport of markers such as horseradish peroxidase enzyme, or fluorescent beads, and they have the advantage of providing an instantaneous snapshot in which all the neurones projecting to a particular area are highlighted. Obtaining a comparable picture using antidromic identification would require multiple experiments and would be much more laborious.

As well as identification of a neurone's project, antidromic stimulation can also be used for studying the electrophysiological features of the recorded neurone. One example is the use of trains of antidromic stimuli to investigate the refractory characteristics of the soma-dendritic spike, as well as the occurrence of hyperpolarizing or depolarizing after-potentials (Zhang *et al.*, 1991). Antidromic stimulation can also reveal the time delay between occurrence of an action potential in the initial segment of the axon and activation of the soma-dendritic spike. Other uses of antidromic stimulation include investigation of the significance of neuronal interactions in the generation of synchronous bursting activity (Negoro *et al.*, 1985), and the regulation of dendritic release from peptidergic neurones (Ludwig *et al.*, 2002).

Further reading and resources

Inyushkin, A.N., Orlans, H.O. & Dyball, R.E. (2009). Secretory cells of the supraoptic nucleus have central as well as neurohypophysial projections. *Journal of Anatomy* **215**, 425–434.

Lincoln, D.W. & Wakerley, J.B. (1974). Electrophysiological evidence for the activation of supraoptic neurones during the release of oxytocin. *Journal of Physiology* **242**, 533–554.

Lipski, J. (1981). Antidromic activation of neurones as an analytic tool in the study of the central nervous system. *Journal of Neuroscience Methods* **4**, 1–32.

Ludwig, M., Sabatier, N., Bull, P.M., Landgraf, R., Dayanithi, G. & Leng, G. (2002). Intracellular calcium stores regulate activity-dependent neuropeptide release from dendrites. *Nature* **418**, 85–89.

Negoro, H., Uchide, K., Honda, K. & Higuchi, T. (1985). Facilitatory effect of antidromic stimulation on milk ejection-related activation of oxytocin neurons during suckling in the rat. *Neuroscience Letters* **59**, 21–25.

Swadlow, H.A. (1998). Neocortical efferent neurons with very slowly conducting axons: strategies for reliable antidromic identification. *Journal of Neuroscience Methods* **79**, 131–141.

Zhang, D.X., Owens, C.M. & Willis, W.D. (1991). Intracellular study of electrophysiological features of primate spinothalamic tract neurons and their responses to afferent inputs. *Journal of Neurophysiology* **65**, 1554–1566.

31
Event-Triggered Averaging, Including Spike-Triggered Averaging

Richard Apps

Physiology & Pharmacology, University of Bristol, UK

31.1   Basic 'how-to-do' and 'why-do' section

Averaging any small but consistent response signal is a powerful approach that is able to increase substantially the signal-to-noise ratio (see below) and so greatly increase the resolution for detection. Take, for example, an experiment in which sensory stimuli are repeatedly delivered to a subject, evoking a consistent electromyographic response from a skeletal muscle. As the responses to repetitive stimuli are recorded, the average time delay (latency; milliseconds) and the average size (millivolts) are progressively built up. This is an example of averaging in both the time and the voltage '*domains*'. The procedure reveals what one might call the 'typical' latency and the 'typical' amplitude of the response. It is a form of '*event-triggered*' averaging, in which the 'event' is the stimulus – what emerges from the averaging are the components of the response that are consistently linked to the stimulus.

Note that an important limitation of the approach is a phenomenon known as '*non-stationarity*' – that is, if the response is variable in latency, size or shape on a trial by trial basis over time, then this variability (which may be biologically important) is not evident in the average. One way to check whether non-stationarity is present is to plot the data on a trial by trial basis and look for any systematic changes over time, for example whether there is any progressive change in response size from first to last stimulus trial.

Despite this potential limitation, averaging is very widely used, not least because when responses are small, it facilitates their detection. This is readily

Essential Guide to Reading Biomedical Papers: Recognising and Interpreting Best Practice, First Edition.
Edited by Phil Langton.
© 2013 by John Wiley & Sons, Ltd. Published 2013 by John Wiley & Sons, Ltd.

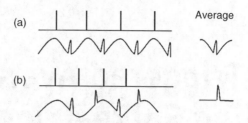

Figure 31.1 The principle of stimulating in antiphase with a regular source of noise. (a) Top trace, stimulus marker; bottom trace, evoked response contaminated with mains interference (50 Hz). Stimulus is delivered in phase with the oscillations of the noise, i.e. tends to occur at the peaks in the oscillations. Averaging results in a response with high levels of background noise. (b) Same as (a), but stimulus is delivered antiphase with the noise, i.e. timed to occur at the peaks and troughs in the oscillations. Averaging therefore tends to cancel out the interference, resulting in a response with lower levels of background noise. Courtesy of Professor Richard Apps.

understandable if the responses are recorded against a background of activity that is not '[1]*time-locked*' to the stimulus. This activity will differ from trial to trial, whereas the response will have components that do not so differ. When averaging of a number of trials is carried out, these response components will sum up from trial to trial, whereas the background activity will tend towards constant levels throughout the duration of the trace.

Regarding the background activity as '*noise*', we can say that averaging has improved the '*signal-to-noise ratio*' (where the '*signal*' is the response to the stimulus). For example, if you wish to deliver a stimulus once every second (i.e. every 1,000 ms), it is possible to reduce the effects of AC electrical mains noise on your recording (which has a frequency of 50 Hz – i.e. a periodicity of 20 ms) by delivering the stimulus at an interval of once every 1,010 ms. This is because the sinusoidal noise will be antiphase for each stimulus delivery, and will therefore tend to cancel out with averaging (see Figure 31.1).

When we consider that biological responses often involve very small voltages (i.e. small signals), and that they are often superimposed on lots of ongoing but unrelated activity (noise), then the utility of the approach is obvious.

So far, the averaging referred to has been 'stimulus-triggered', but other events can be used as triggers. For example, a sudden change in one body system (system A) can be used as the trigger for averaging activity in another system (system B), thereby revealing the activity pattern that consistently follows the chosen trigger event. This pattern is evident in a chart called a '*peri-event*' or '*peri-stimulus*' time histogram (PETH or PSTH). It is important to note that the approach does not demonstrate that the trigger event in A *causes* the change in system B; they may only be correlated. It might be, for example, that some third event not known

---

[1] Background activity that is *not* time-locked has no observable relation with the timing of the stimulus.

to the experimenter might cause both the trigger event in A and the change in B (see pitfalls below and Primers 2 and 3).

One form of averaging that is a very powerful tool in systems neuroscience is 'spike-triggered averaging' (STA). In this approach, a train of action potentials (spikes) is recorded from a single neurone (neurone X) and, simultaneously, the spike train of another neurone (neurone Y) is also recorded. If neurone X supplies one or more excitatory synapses to neurone Y, then each time X fires an action potential, there will be, after an impulse conduction and a synaptic delay, an excitatory post-synaptic potential (EPSP) recorded in Y. If each spike of X is used as a trigger for averaging the activity of Y (i.e. a 'spike-triggered' average or STA is recorded), then the influence of X on Y can be revealed. The relation that emerges is plotted as a cross-correlogram (see Figure 31.2).

Note that if X does not connect with Y, then the averaging will reveal nothing – unless, of course, there is a third neurone which consistently influences both X and Y.

The activity of Y that is averaged can be:

- the membrane potential – in which case, the averaging will reveal the shape and size of the EPSP.

- the action potentials fired by Y. In this case, firing rate of Y is plotted against time elapsed since each trigger spike in X.

There is an important point and principle to recognize here. If a spike in X gives rise to an EPSP in Y, then, however small that EPSP might be, its presence will slightly increase the probability of firing in Y. If Y has an underlying pattern of firing then the plot of firing rate versus time will reveal an EPSP-shaped (i.e. mirroring both amplitude and duration) increase in the firing of Y. The increase is often called a 'post-spike' facilitation (PSF – this means a facilitation of Y after each X spike).

Spike-triggered averaging can also work if the connection from X to Y is inhibitory, in which case the inhibitory post-synaptic potential (IPSP) can be revealed either directly (by recording membrane potential of Y) or indirectly (via a reduction in firing rate).

Note also that one special case of spike-triggered averaging is possible for alpha motoneurones. If the aim is to see whether a particular neurone synapses with an alpha motoneurone, there is no need to record directly from the motoneurone. Each time it fires, its motor unit muscle fibres will develop an action potential, so records can be made of these (i.e. a single unit EMG record can be made and used as the Y record). The advantage here is that EMG records are much easier to make than recording the spike trains of an individual neurone, particularly over the periods of time necessary to generate a spike triggered average (see pitfalls).

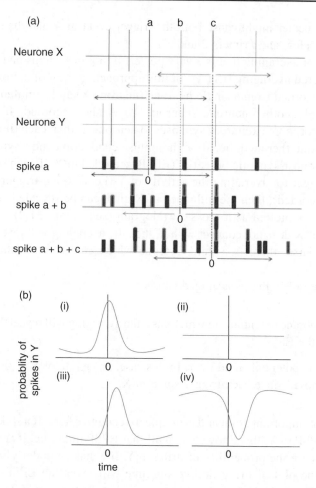

Figure 31.2 Aspects of analysis of spike-triggered averaging – the cross-correlogram (a) Construction of a cross-correlogram. Spike trains of two neurones recorded simultaneously. Time of occurrence of each spike in X is taken as time zero, and the time of spikes in Y are plotted on a histogram in relation to each time zero marker. Construction of a peri-event histogram is shown in relation to time of three spikes in X (spikes a, b and c). In this example, there is an increase in frequency of occurrence of spikes in Y just after a spike occurs in X. This is consistent with a monosynaptic excitatory connection between X and Y. (b) Examples of different shapes of cross-correlogram. (i) X and Y are synchronously active; (ii) activity of X and Y are not correlated; (iii) an increase in probability of spikes in Y occurs after activity in X, consistent with X exciting Y; (iv) a reduction in probability of spikes in Y occurs after activity in X, consistent with X inhibiting Y. Time zero = occurrence of spikes in X. Courtesy of Professor Richard Apps. *A full colour version of this figure appears in the colour plate section.*

31.2   Pitfalls in execution or interpretation

- Though spike-triggered averaging is a very valuable way of showing whether neurone X connects monosynaptically with neurone Y, it is necessary to keep in mind that false positives can occur if a third neurone projects to both X and Y, as having this input in common can mimic an XY connection. A related problem is that correlated activity can occur between X and Y if the firing rates of the two cells co-vary, for example if their activity is time-locked to a stimulus, or if the excitability of the two cells fluctuates in parallel (Brody, 1999).

- The effect of X on Y is also almost always very small (because EPSPs and IPSPs are small), so it is usually necessary to record thousands of impulses from both neurones (and the recordings must, of course, be made for the same time). This is technically challenging.

- In theory, the connection from X to Y could also involve more than one synapse (i.e. there could be interneurones) in the pathway. However, in practice, spike-triggered averaging only really works for monosynaptic connections. If interneurones are present, they introduce so much extra variation or 'noise' that no clear-cut post spike effect can usually be uncovered, unless the polysynaptic pathway is particularly secure.

- Care should also be taken when comparisons are made between activity of cells when their firing rates change under different behavioural conditions. For example, an increase in firing frequency will result in an increased tendency for the two cells to fire at the same time. The choice of bin width in the analysis is also an important consideration – clearly, the wider the time bins used to construct the histogram, the greater the opportunity for overlaps in time to occur. Bin widths between 1–10 ms are typically used, but they depend on firing frequency of the cells under study and the total number of spike events recorded. Cross-correlogram analysis programs that take these various issues into account are available.

31.3   Complementary and/or adjunct techniques

Primer 28: Electrical stimulation methods.
Primer 29: Extracellular recording.
Primer 30: Antidromic identification.
Primer 32: Axonal transport tracing of CNS pathways.

Further reading and resources

Fetz, E. & Cheney, P. (1980). Postspike facilitation of forelimb muscle activity by primate cortico-motoneuronal cells. *Journal of Neurophysiology* **44**, 751–752.

Blenkinsop, T.A. & Lang, E.J. (2011). Synaptic action of the olivocerebellar system on cerebellar nuclear spike activity. Journal of Neuroscience **31**(41), 14708–14720.

Brody, C.D. (1999). Correlations without synchrony. Neural Computation **11**(7), 1537–1551.

Fetz, E. & Cheney, P. (1980). Postspike facilitation of forelimb muscle activity by primate cortico-motoneuronal cells. Journal of Neurophysiology **44**, 751–752.

Palm, G., Aertsen, A.M. & Gerstein, G.L. (1988). On the significance of correlations among neuronal spike trains. Biological Cybernetics **r59**, 1–11.

32
Axonal Transport Tracing of CNS Pathways

John Crabtree
Physiology & Pharmacology, University of Bristol, UK

32.1   Basic 'how-to-do' and 'why-do' section

Axonal transport tracing of pathways defines connections between structures in the central nervous system (CNS). These CNS structures are [1]*aggregations* of neurons that usually have well-defined borders, can be visualized by histology, and are usually referred to as nuclei, ganglia, laminae or layers. Within these aggregates are projection neurons that send their axons outside an aggregate to another structure.

CNS structures are often reciprocally connected and can form complex networks of connectivity both within and across levels of the neural axis. Unravelling this connectivity is the goal of pathway tracing, which will provide a fundamental structural framework upon which to build our understanding of the function of the CNS.

In the late 19th century, Golgi preparations (staining of neurons with silver chromate) allowed investigators to visualize for the first time individual axons and their branches in their entirety, enabling a rough framework of connectivity for the CNS (Figure 32.1).

In the second half of the 20th century, CNS pathway tracing techniques were developed that use the axoplasmic transport systems of neurons and allow:

- anterograde (forward) tracing, in which a tracer is taken up by the cell body and is transported down the axon to the terminals;

- retrograde (backward) tracing, in which a tracer is taken up by terminals and transported up the axon to the cell body;

- both anterograde and retrograde tracing.

---

[1] **Aggregation** – meaning: group, cluster, collection.

Essential Guide to Reading Biomedical Papers: Recognising and Interpreting Best Practice, First Edition.
Edited by Phil Langton.
© 2013 by John Wiley & Sons, Ltd. Published 2013 by John Wiley & Sons, Ltd.

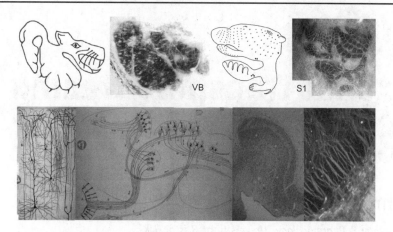

Figure 32.1   Visualizing neurons in the CNS. **Upper left panels**: Monoclonal antibody Cat-301 is directed against the cell-surface proteoglycan chondroitin sulphate of neurons and targets specific structures in the CNS. The section on the right shows an example of Cat-301 immunoreactivity in the cat. The section was taken in the oblique coronal plane through the first-order somatosensory nucleus of the thalamus, the ventrobasal complex (VB). Cat-301 stains neurons in the entire VB giving this structure a compartmentalized appearance. Rostral is at the top and medial is to the right. Based on receptive field mapping studies, the drawing on the left is a caricature of the cat body surface and reflects the compartments of the Cat-301 neuronal stain shown on the right. Figure 5 within Crabtree JW. (1999). Intrathalamic sensory connections mediated by the thalamic reticular nucleus. Cell Mol Life Sci. 1999 Nov 15;56(7–8):683–700. © Springer Publishing. **Upper right panels**: Example of staining for the mitochondrial enzyme succinic dehydrogenase in the rat. The section on the right was taken tangential to the cerebral cortex through layer 4 of the flattened primary somatosensory cortex (S1). Lateral is at the top and rostral is to the left. Based on receptive field mapping studies, the drawing on the left is a caricature of the rat body surface and reflects the morphological pattern shown on the right. Reproduced, with permission, from Dawson DR, Killackey HP. (1987). The organization and mutability of the forepaw and hindpaw representations in the somatosensory cortex of the neonatal rat. J Comp Neurol. 1987 Feb 8;256(2):246–56. © John Wiley & Sons Ltd. **Lower left panels**: Drawings by Ramón y Cajal based on observations from Golgi-stained preparations. The drawing on the left shows 'First, second and third layers of the precentral of the cerebrum of a child of one month'. Projected onto the coronal plane, details of the cerebral cortex including dendritic spines are shown. The drawing on the right shows 'Scheme of the afferent and efferent pathways of the optic centers'. Projected onto the sagittal plane, connections of the retina (A) to the dorsal lateral geniculate nucleus (dLGN; D) and the superior colliculus (E) and the projection from the dLGN to the visual cortex (G) are shown. Drawings reproduced with the permission of the Inheritors of Santiago Ramón y Cajal ©. **Near right lower panel**: Nissl stain of the first-order visual nucleus of the thalamus, the dorsal lateral geniculate nucleus, of a cat (Crabtree, unpublished). Two laminae of neurons are clearly seen in the sagittal plane, an outer A lamina receiving inputs from the contralateral retina and an inner A1 lamina receiving inputs from the ipsilateral retina. Dorsal is at the top and rostral is to the right. **Far right lower panel**. Example of visualizing transparent brain tissue by using only the properties of light, that is, without using stains (Crabtree, unpublished). The section was taken in the horizontal plane through the internal capsule at the level of the thalamus in the cat. Two light sources were shone onto the section – one from above and one from below – producing contrast interference patterns and revealing large stringy bundles of axons (fasciculi) in the internal capsule (lower left to upper right) that de-fasciculate (upper left) as the axons approach the cerebral cortex. It is important to note that the techniques used to trace pathways in the CNS are generally too fine-grained to reveal large organizational features in the CNS, such as the maps shown in the upper panels. Courtesy of Dr. John Crabtree. *A full colour version of this figure appears in the colour plate section.*

The first tracers made available were tritiated (radioactive) amino acids and the enzyme horseradish peroxidase (HRP), later joined by the plant lectins wheat germ agglutinin (WGA) and *Phaseolus vulgaris*-leucoagglutinin as well as biotinylated dextran, biocytin, and cholera toxin. Fluorescent tracers were also introduced including latex microspheres, nuclear yellow, fast blue, and carbocyanine dyes. Some of these tracers can be coupled together (e.g., WGA-HRP). A general protocol for CNS pathway tracing is outlined below, but in practice there will be differences among the protocols for each tracer used (Bancroft & Gamble, 2002; Crabtree, 1998).

Step 1   The animal is deeply anaesthetized and mounted in a stereotaxic apparatus. A trephine hole is made in the skull over a target structure. Using pulses of pressure, a tracer is injected through a glass micropipette or syringe needle into the target structure according to its[2] *stereotaxic coordinates.* The concentration of the solution injected will depend upon the tracer used and the volume injected will depend upon the size of the target structure.

Step 2   An appropriate post-injection interval must be observed to allow the tracer to be transported to the terminal destination of cell bodies at the injection site (anterograde transport), or to the cellular origin of axon terminals at the injection site (retrograde transport), or to both. The post-injection interval will depend upon the speed at which the tracer is axoplasmically transported and the estimated distance that must be traversed.

Step 3   The animal is deeply anaesthetized and perfused transcardially with a mixture of paraformaldehyde and glutaraldehyde. The relative proportions of these fixatives will depend on the tracer used. Glutaraldehyde should be avoided altogether if a fluorescent tracer is used, as glutaraldehyde is itself fluorescent. The fixed brain is removed and can be immersed in aldehydes for further fixation (12 hours or longer) but, eventually, the brain must be dehydrated by immersing it in a sucrose solution (a 30 per cent solution is often used) for 12–48 hours depending on the size of the brain.

Step 4   Once dehydrated, the brain is frozen and cut into sections (commonly 50 μm thick). Popular planes of section are coronal, horizontal, and sagittal. However, there are no golden rules dictating the plane in which a brain should be cut; any plane is allowed, provided it is clearly defined so that others can repeat the study if necessary. The important point is to cut the brain in a plane that will optimize the connectivity that you expect to see – or, if uncertain, cut different brains in different planes. Key organizational features of the brain can be obscured simply by choosing the wrong plane of section!

---

[2] **Stereotaxic coordinates** – meaning: reference to a three-dimensional atlas of the brain, for example, the atlas of Paxinos & Watson, 1998.

Step 5   The brain sections may or may not require histological processing to visualize the tracer (now referred to as 'labelling') in its final position at the time of brain fixation. When using many of the fluorescent tracers, sections are simply mounted on slides and the labelling viewed by fluorescence microscopy (Primer 8). However, a chromogen, or colouring agent, must be deposited at sites containing non-fluorescent tracers so that their locations can be visualized. For example, one such chromogen is tetramethyl benzidine, which provides a blue-black reaction product at sites containing HRP. Once a chromogen has been added, the sections are mounted on slides for microscopic analysis.

Step 6   Finally, distributions of the labelling are reconstructed. Labelling is visualized in at least two locations – the site of the tracer injection and the region(s) to which the tracer was transported. The distributions of labelling are invariably computer imaged and/or photographed for illustrative purposes, and they can also be drawn. Drawings include the locations of major blood vessels as reference points and the borders of neuronal aggregates (nuclei, ganglia, laminae, layers) after Nissl staining (see controls below). Any quantitative analysis (e.g., cell sizes, cell numbers) of labelling can be made directly from the sections on the slides under microscopic viewing.

32.2   Required controls

1. The validity of any interpretation of the results of a pathway tracing study will largely depend on a computer imaging and/or photographic demonstration of the precise locations and extents of the injection site and the region(s) to which the tracer was transported. Line drawings of such are no longer acceptable as these are prone to flights of fancy (i.e. observer bias – see primer 2). To determine these locations and extents, either preselected sections through the injection site and expected transport site(s) must be set aside for Nissl staining or sections containing labelling must be Nissl counterstained. Nissl stains (e.g. cresyl violet, thionin, neutral red) stain basophilic granules representing rough endoplasmic reticulum in the cytoplasm and (hopefully) clearly reveal the neuronal borders of nuclei, ganglia, laminae, or layers (Figures 32.1 and 32.2).

2. Physiological identification of neurons (e.g. by their response properties) at an injection site is necessary if the borders containing the neurons are poorly defined.

3. Because tracers vary in their propensity for uptake, axoplasmic transport properties, and sensitivity for detection, replicating work using a different pathway tracer is often desirable.

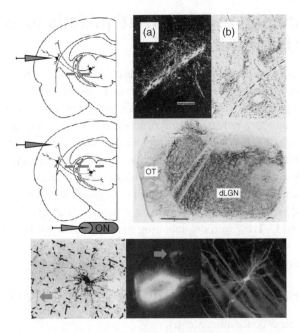

Figure 32.2 Pathway tracing in the CNS. **Top panels:** Example of anterograde tracing using a tritiated amino acid in the cat (from Crabtree 1992). The schema on the left shows the experimental setup projected onto the coronal plane. Injection of a tracer was made in layer 6 of the cerebral cortex, the neurons of which project to the thalamic reticular nucleus (R) and thalamic relay nuclei. The dashed line shows where sections were taken in the horizontal plane. (a, b): sections through R. Rostral is at the top and medial is to the right. (a): section showing silver grains following an injection of [3H]proline made in the cortex (dark-field optics). The calibration bar is 200 μm. (b): same section and field of view (Nissl stain). The dashed line indicates the anterolateral border of a thalamic relay nucleus. (left) Reproduced, with permission, from Crabtree JW. (1992). The Somatotopic Organization Within the Cat's Thalamic Reticular Nucleus. Eur J Neurosci. 4(12): 1352–1361. © Wiley. (right) Courtesy of Dr. John Crabtree. **Middle panels:** Examples of anterograde and retrograde tracing using the enzyme horseradish peroxidase (HRP) in the rabbit (Crabtree, unpublished). The schema on the left shows the experimental setup projected onto the coronal plane. Injections of HRP were made in the contralateral optic nerve (ON) and in layer 4 of visual cortex, which contains terminals of the axonal projections from the dorsal lateral geniculate nucleus (dLGN) of the thalamus. The dashed line shows where sections were taken in the horizontal plane. A section showing the resultant HRP labelling is shown on the right. Almost the entire dLGN is filled with anterograde labelling of terminals where HRP was transported via the axons of the ON and optic tract (OT). A small label-free region representing input from the ipsilateral ON can be seen on the far right. The dLGN also contains a column of retrograde labelling of cell bodies (insert) where HRP was transported via the axons of dLGN relay cells. Rostral is at the top and medial is to the right. The calibration bar is 500 μm. **Lower left panel:** Example of a biocytin-filled neuron in a thalamic relay nucleus of the rat (from Crabtree JW, Isaac JT. (2002). Reproduced, with permission, from Crabtree JW. (1992). The Somatotopic Organization Within the Cat's Thalamic Reticular Nucleus. Eur J Neurosci. 4(12): 1352–1361. © Wiley. New intrathalamic pathways allowing modality-related and cross-modality switching in the dorsal thalamus. J Neurosci. 2002 Oct 1;22(19):8754–61. © Society for Neuroscience). Intracellular injection of biocytin gives neurons a Golgi-like appearance. This allows visualization not only of the trajectory and extent of single axons and their branches but also of morphological features such as the size and shape of the neuronal soma and dendritic tree. The arrow points to the axon of the biocytin-filled neuron. **Lower right panels:** Example of retrograde

(continued on next page)

32.3   Common problems and pitfalls in execution or interpretation

Although we have at our disposal a large and varied assortment of pathway tracers, no one tracer is entirely problem free. To various extents, use of a tracer can result in what we will lump together as *false negatives* or *false positives*. For example, we make a small injection of a retrograde tracer into structure B but see no retrograde labelling in structure A, concluding that A does not project to B. However, our lack of labelling is a false negative, because the volume of tracer injected was too small and too localized to be taken up adequately by the sparse and widespread terminals from A in B.

Consider another example. We make an injection of an anterograde tracer into structure A and see anterograde labelling in structures B and C, concluding that A projects to B and C. However, our labelling in C is a false positive, because axons merely passing through A on their way to C were damaged and took up the tracer. This latter example is called the '*fibres-of-passage*' problem.

32.4   Complementary and/or adjunct techniques

primer 8: Fluorescence microscopy.

primer 28: Electrical stimulation methods.

primer 30: Antidromic identification.

Further reading and resources

Bancroft, J.D. & Gamble, M. (Eds.) (2002). *Theory and practice of histological techniques*, 5th Edition, Harcourt Publishers Limited.

Cowan, W.M. (1998). The emergence of modern neuroanatomy and developmental neuro-biology. *Neuron* **20**, 413–426.

Crabtree, J.W. (1998). Organization in the auditory sector of the cat's thalamic reticular nucleus. *Journal of Comparative Neurology* **390**, 167–182.

Crabtree, J.W. (1999). Intrathalamic sensory connections mediated by the thalamic reticular nucleus (review). *Cellular and Molecular Life Sciences* **56**, 683–700.

Paxinos, G. & Watson, C. (1998). The Rat Brain in Stereotaxic Coordinates. Academic Press, San Diego, CA.

◀ ───

(*Continued*)

tracing using the fluorescent carbocyanine dye DiI in the rabbit (from Crabtree 1999). Two sections taken in the coronal plane through the thalamus are shown. On the left, labelling is shown after a small crystal of DiI was placed in the dorsal part of the thalamus (low magnification). The arrow points to a cluster of labelled neurons in the thalamic reticular nucleus. On the right, a higher magnification of one such labelled neuron is shown against a background of labelled axons. Unlike other anterograde and/or retrograde tracers, the carbocyanine dyes do not use the neuronal axoplasmic transport systems; rather, they diffuse along the extracellular surface of a neuron and its processes. It is important to note that the techniques used to trace pathways in the CNS are generally too fine-grained to reveal large organizational features in the CNS, such as the maps shown in the upper panels of Figure 32.1. Reproduced, with permission, from Crabtree JW. (1999). Intrathalamic sensory connections mediated by the thalamic reticular nucleus. Cell Mol Life Sci. 1999 Nov 15;56(7–8): 683–700. © Springer. *A full colour version of this figure appears in the colour plate section.*

33
Cardiovascular Methods: General Considerations for Human Studies

Erica A Wehrwein[1] and Michael J Joyner[2]

[1]*Physiology, Mayo Clinic, Minnesota, USA*
[2]*Anesthesia Research, Mayo Clinic, Minnesota, USA*

33.1 Basic 'how-to-do' and 'why-do' section

It should be obvious that research involving human subjects is an essential component to understanding the physiology of the human cardiovascular system. Human subject research is necessary to complement the investigations done in cellular and animal models, and not only in the case of clinical trials research; studies in humans are also important for understanding basic systems physiology. This is especially true for the multiple highly integrated control mechanisms involved in mediating cardiac function, blood pressure and blood flow. The considerations below are illustrated with examples from the cardiovascular research field, but the underpinning principles are entirely generic.

The Declaration of Helsinki clearly states, in paragraphs 21 and 22, that:

> *'Medical research involving human subjects may only be conducted if the importance of the objective outweighs the inherent risks and burdens to the research subjects.*
>
> *Participation by competent individuals as subjects in medical research must be voluntary. Although it may be appropriate to consult family members or community leaders, no competent individual may be enrolled in a research study unless he or she freely agrees.'*

Essential Guide to Reading Biomedical Papers: Recognising and Interpreting Best Practice, First Edition.
Edited by Phil Langton.
© 2013 by John Wiley & Sons, Ltd. Published 2013 by John Wiley & Sons, Ltd.

Given these principles, there are two fundamental considerations that are of relevance to the design of experiments using human subjects:

1. *Variability in human subjects:* Unlike cellular or animal studies, which use a genetically identical (or highly similar) line of cells or inbred animal strain, humans are very diverse. It is important to understand that the study will have been designed in such a way that the variability is limited as much as possible (see below under 'required controls').

2. *Limited instrumentation and indirect measurements:* In many cases, laboratories do not have the expertise or resources to perform highly invasive studies in which subjects have monitoring that involves placement of catheters, biopsies, internal monitoring of organ function, etc. In addition, it may not be deemed ethical by an institution to do such studies if there are suitable non-invasive options available. Therefore, in many cases, researchers must rely upon non-invasive or minimally invasive techniques as an alternative. It is important to note when reading these studies that, although an invasive technique may seem like a better alternative, a non-invasive or indirect measurement that has been validated can often be used reliably, assuming that it is properly controlled.

33.1.1   Temporal considerations

Depending on the outcome of interest, researchers will choose the appropriate time frame for data collection. In broad terms, there are both *acute* and *chronic* recordings, each offering a different type of data.

In the *acute* category, there are commonly studies that involve a single observation in a laboratory setting. In these instances, a research subject would have a single visit to the laboratory for a one-time assessment. These one-time assessments can involve a single time point of data, repeated measurements of the same parameter over a short duration for averaging, or continuous monitoring in laboratory for the duration of an intervention which ranges from several minutes to several hours. It is important to consider that the acute measurements, even those that use short-term continuous monitoring, only offer a 'snapshot' of physiology at that time under the conditions studied. Therefore, it is possible that this would not offer a complete picture, and it may be that a repeated measurement on a different day, or collected in the long term, would reveal different results in the same person.

In the *chronic* category, a research subject can be instrumented (e.g. non-invasively with a portable blood pressure cuff) for remote monitoring of physiological parameters of interest, or there can be a long-term treatment that is monitored by repeated visits to the laboratory over time. The former can be done with telemetric monitoring, where the data is downloaded continuously, or by a separate recording device that saves data over a time (e.g. a 24 hour monitoring period), to be downloaded and analyzed later. Beyond this, there are also longitudinal studies that will monitor subjects repeatedly over many years.

33.2 Required controls

For studies in the United States, the U.S. Department of Health and Human Services issued human subject protection regulations in 1974. These were followed by the Belmont Report in 1978, describing the ethical principles and guidelines for human subjects research.

When reviewing current papers involving human subjects, there are several general things to watch for in the 'methods' description. First and foremost, every study involving human subjects is required to meet strict ethical guidelines for study design and protection of the rights of the individuals involved in the study. Therefore, there needs to be a statement of approval for the study from the institution where the study was conducted, including that the subjects provided written consent to participate. In order to provide consent, a subject must be fully informed of the pros and cons of participating, and any potential risk(s) that they may encounter. The data collection itself must strictly follow these guidelines as approved both for the safety of the research participants and for the integrity of the data collection.

As stated above, it is essential that studies are tightly controlled such that experimental conditions are precisely matched across subjects (see Primers 2 and 3). That is to say, studies may be limited to a particular age, physical activity level, race, gender, body mass index, disease condition, etc. This is especially important when comparing a control to experimental group. It is essential to 'match' subjects properly in both groups, on as many of the above parameters as possible, to limit variability. To ensure that the data observed are valid, it is also common to perform studies at a particular time of day, at a certain phase of the menstrual cycle, in the absence of confounding medications, often while subjects are fasting, and/or after a standardized meal to further ensure consistency for all subjects. Depending on the outcome of interest, it may be also necessary to control for other parameters, including but not limited to, caffeine intake, hydration status, room temperature/humidity, time of day, etc.

In broad terms, the methods for studying the highly integrative human physiological system can be thought of in terms of a music analogy. Let us consider that the methods used to assess components of human cardiovascular function are like playing a musical instrument. In this analogy, it is vital that the instrument is fully functional and properly tuned (not unlike the equipment used to collect the data being properly maintained and calibrated). The musician needs to have sufficient practice playing the instrument as to easily recognize when it is out of tune (similarly, it is typical that human researchers will check calibration with each use and will practice data collection often, so that they thoroughly know and understand the equipment and when it may be malfunctioning).

With much practice, the musician knows the music very well and can quickly recognize when an incorrect note is played (during real-time human data collection, it is possible that an acute issue will arise with the subject or the equipment, and it is vital to realize this and correct it as soon as possible). Finally, and importantly, the musician (researcher) must understand where their part fits into the whole performance (e.g. how does the data for blood flow velocity correspond to blood pressure and other systemic functions?).

Being mindful of all of these aspects is important for the musician during practice and performance in order to produce a beautiful piece of music. Likewise, the researcher needs to be mindful of the equipment, subject, and data, both while the data is being collected and then when it is being analyzed later.

These general themes – having a tuned instrument, knowing the music and understanding how each instrument fits into the whole performance – are reflected in discussion of controls and common problems in the primers on human methods measuring cardiac output and blood flow (see Primers 34 and 35).

33.3   Common problems or errors in literature and pitfalls in execution or interpretation

- *Small sample size.* The numbers of subjects can be small, but it is practically impossible that the study will have been approved by the Ethics Board if the statistical power of the study was not known to be favourable. Statistical power, however, cannot be determined without some knowledge of the variability in the measured variables and the size of the expected treatment effect. Without pilot experiments, it is often difficult to acquire the data necessary to calculate the size of the groups.

- *Lack of consideration for gender differences in the parameter of interest.* Unless there is an obvious gender specificity in the experiments, the experimental groups should include both males and females, this should also be part of the matching strategy. It would be expected that the results would be analyzed by gender at first, combining the data only if there is no evidence of gender difference. This approach will have been taken into account when considering the size of the experimental groups.

- *Lack of control for menstrual cycle phase in female subjects.* There is now good evidence that female sex hormones influence the heart, sympathetic nervous system and cardiovascular disease. In order that variability in the data is minimized, measurements should be scheduled such that the phase of the menstrual cycle is not a factor.

- *Lack of control of physical activity level and overall fitness.* Cardiovascular physiology is very powerfully influenced by physical activity. This is now beginning to be included in the list of factors to be matched across experimental groups.

- *Drawing conclusions beyond the scope of the study* (e.g. extending findings in young women to older women). This is an issue in all areas of science. It is sensible to ask the question, 'does the same hold true for group X?' but it is not sensible to make that assumption.

- *Extending the conclusions drawn under the particular study design to other similar study designs that may have key differences* (e.g. studies on regulation of blood flow in a particular vascular bed are not necessarily applicable to other vascular beds). Animals studies have provided clear evidence that different vascular beds in the same individual animal can exhibit quite marked differences in their behaviour and their pharmacology. It is wise to assume that this observation will be upheld in human studies.

- *Use of body mass index as a screening tool or matching parameter.* Body mass index is often used as a convenient way to match subjects and to screen potential subjects. However, in and of itself, the body mass index does not necessarily provide good information on an individual basis in terms of actual body composition. In other words, body mass index gives a generally understanding of the body weight per height, but does not tell us anything specific about lean body mass or body fat percentage. When it is important to have a good understanding of lean body mass (e.g. when studying insulin sensitivity), then further testing, such as dual-energy X-ray absorptiometry (DEXA), should also be included.

- *Studies that involve aging populations.* There is substantial literature regarding age-specific changes in many systems, which raises particular issues of how to best study aging. One approach is to try to match an aging individual to a young, healthy individual on as many parameters as possible. Issues arise in defining age categories (i.e. at what age is one considered 'old' or 'very old'?) and often, females are categorized as being pre- or post-menopausal. It is also difficult to recruit truly healthy aging individuals that are physically active and do not take medications. Since activity and medications are confounding variables, these are important factors when trying to match subjects and when attempting to study healthy aging.

33.4 Complementary and/or adjunct techniques

Primer 1: Philosophy of science.
Primer 2: Experimental design.
Primer 3: Statistics.
Primer 34: Cardiac output – human studies.
Primer 35: Peripheral blood flow – human studies.

Further reading and resources

Guidelines for association studies in human molecular genetics: http://hmg.oxfordjournals. org/content/14/17/2481.full

The Belmont Report: Ethical principles and guidelines for the protection of human subjects of research: http://ohsr.od.nih.gov/guidelines/belmont.html

WMA Declaration of Helsinki – Ethical principles for medical research involving human subjects: www.wma.net/en/30publications/10policies/b3/

The following paragraphs are taken from the Declaration of Helsinki:

'1. The World Medical Association (WMA) has developed the Declaration of Helsinki as a statement of ethical principles for medical research involving human subjects, including research on identifiable human material and data. The Declaration is intended to be read as a whole and each of its constituent paragraphs should not be applied without consideration of all other relevant paragraphs.

12. Medical research involving human subjects must conform to generally accepted scientific principles, be based on a thorough knowledge of the scientific literature, other relevant sources of information, and adequate laboratory and, as appropriate, animal experimentation

14. The design and performance of each research study involving human subjects must be clearly described in a research protocol. The protocol should contain a statement of the ethical considerations involved and should indicate how the principles in this Declaration have been addressed. The protocol should include information regarding funding, sponsors, institutional affiliations, other potential conflicts of interest, incentives for subjects and provisions for treating and/or compensating subjects who are harmed as a consequence of participation in the research study. The protocol should describe arrangements for post-study access by study subjects to interventions identified as beneficial in the study or access to other appropriate care or benefits.

15. The research protocol must be submitted for consideration, comment, guidance and approval to a research ethics committee before the study begins. This committee must be independent of the researcher, the sponsor and any other undue influence. It must take into consideration the laws and regulations of the country or countries in which the research is to be performed as well as applicable international norms and standards but these must not be allowed to reduce or eliminate any of the protections for research subjects set forth in this Declaration. The committee must have the right to monitor ongoing studies. The researcher must provide monitoring information to the committee, especially information about any serious adverse events. No change to the protocol may be made without consideration and approval by the committee.

21. Medical research involving human subjects may only be conducted if the importance of the objective outweighs the inherent risks and burdens to the research subjects.

22. Participation by competent individuals as subjects in medical research must be voluntary. Although it may be appropriate to consult family members or community leaders, no competent individual may be enrolled in a research study unless he or she freely agrees.'

34
Measuring Cardiac Output in Humans

Erica A Wehrwein[1] and Michael J Joyner[2]

[1]Physiology, Mayo Clinic, Minnesota, USA
[2]Anesthesia Research, Mayo Clinic, Minnesota, USA

34.1   Basic 'how-to-do' and 'why-do' section

There are many features of heart function that are commonly assessed. The most common and simple are heart rate and electrocardiogram. In many cases, it is of interest to determine not only the rate and rhythm of the heart, but also to assess the pumping efficiency and volume of blood expelled with each heartbeat.

Cardiac output (litres of blood pumped by the heart per minute) is the product of heart rate and stroke volume. It gives a valuable measurement of the total blood flow produced by the heart. In healthy adults at rest, cardiac output is approximately 4–6 litres/min. In many disease conditions, however (including hypertension, heart failure and sepsis), cardiac output is altered, so it is important to have an accurate method to determine cardiac output. As mentioned above, the heart rate is routine and simple to assess, but the assessment of stroke volume and cardiac output is more challenging and may be performed invasively (e.g. using intra-cardiac catheterization) or non-invasively (e.g. surface pulse wave interpretation), as described below.

In general terms, the methods to measure cardiac output fall into two categories:

A. Those that rely on principles of mass transport, in this case dilution.

B. Those that are related to solubility and partial pressures.

Methods that use dilution (A) require catheter placement, which is an invasive procedure, while non-invasive techniques (B) are technically less challenging. Both

Essential Guide to Reading Biomedical Papers: Recognising and Interpreting Best Practice, First Edition.
Edited by Phil Langton.
© 2013 by John Wiley & Sons, Ltd. Published 2013 by John Wiley & Sons, Ltd.

these approaches make use of the same experimental logic – the same principle, Fick's Principle.

34.1.1   The Fick principle and measurement of cardiac output

Adolph Fick, in 1870, described a principle that that now bears his name. The logic is simple; the total release or uptake of any given substance from any organ is determined by the product of blood flow through that organ and the difference in the concentrations of that substance measured from arterial and venous blood (Prabhu & Gulve, 2004).

When applied to oxygen as an example, the Fick Principle can be used to calculate how much oxygen is consumed by the body in one minute (VO_2), if one knows the volume of blood pumped by the heart in that time (output in one minute = cardiac output = Q) and can measure the [1]content of oxygen in the venous (C_vO_2) and arterial (C_aO_2) blood. The equation is:

$$VO_2 = Q(C_aO_2 - C_vO_2)$$

The problem, of course, is that Q is the hardest of these variables to measure. In 1887, Fick applied his principle to the problem of measuring cardiac output (Q) which required only a rearrangement of the above equation, making Q the object of the equation (by dividing both sides of the equation by ($C_aO_2 - C_vO_2$)).

$$Q = \left(\frac{VO_2}{C_aO_2 - C_vO_2} \right)$$

34.1.2   Understanding the fick equation: the train analogy

Since the Fick principle is the basis for several methods discussed below, it is useful to clarify this further using the analogy of a cargo train (Grossman, 2006). We will assume that the train, with many open-topped cars, represents the circulation, and that sacks represent the oxygen. The cars receive their cargo of sacks (oxygen) that are dropped in each car as it passes by a hopper (lungs). Since there is already some oxygen in the circulation, we can think of the train cars being partially filled with cargo as they arrive at the hopper. Using this train analogy, we will solve the problem using Fick's Principle and by doing nothing more complex than subtracting or dividing two numbers:

- Each boxcar represents 1 L of blood in the circulation

- Each sack is 1 ml of oxygen picked up by that litre of blood from the lungs.

[1] Oxygen content of blood is measured in ml/litre. Arterial blood will have an oxygen content of about 195 ml O_2/litre, and mixed venous blood around 145 ml O_2/litre.

Let us assume that hopper is delivering sacks at a constant rate of 250 per minute. Let us also assume that we are able to weigh each car as it approaches the hopper and again as it leaves the hopper, and that from the weight we know how many sacks each car contains.

So, approaching the hopper, each train car contains 150 sacks and, after passing the hopper, it contains 200 sacks (admittedly, it is a rather inefficient way to use a train).

This means that each car is picking up 50 sacks from the hopper. We already knew (above) that the hopper is supplying 250 sacks per minute, so it must be that five of the cars pass under the hopper every minute (because $250/50 = 5$). If we convert back from trains to the cardiovascular system:

- Hopper = lungs

- 1 car = 1 litre of blood

- 1 sack = 1 ml of oxygen

A total of 250 ml O_2 are delivered by the lungs *every minute*. Each litre of *venous* blood approaches the lungs containing 150 ml O_2, and each litre of *arterial* blood leaves containing 200 ml O_2. The total volume of blood that must pass through the lungs (Q) is:

$$Q = \frac{250}{(200 - 150)} = \frac{250}{50} = 5 \text{ litres per minute}$$

34.1.3   Data collection to be used for fick calculation

For the above equations, whole body oxygen consumption can be determined using a respiratory analysis device, known as a spirometer, with a mouthpiece or mask used to measure inspired and expired oxygen content along with the minute ventilation (in litres per minute). Alternatively, a metabolic rate meter may be used.

Oxygen content of mixed venous blood is determined from the pulmonary artery as it exits the right atrium en route to the lungs. Blood samples are collected using a thin catheter inserted into a peripheral vein and passed, under X-ray visualization, through the right atrium into the pulmonary artery (this is the invasive aspect). Arterial oxygen content is measured from arterial blood samples obtained from a cannula placed in a peripheral artery. These three values – oxygen consumption, venous oxygen, and arterial oxygen – are used in the equation above to determine cardiac output using the Fick Principle.

The *Fick method* for measuring cardiac output is highly invasive, requires technical expertise and carries a non-trivial risk of infection and/or injury. It requires physician oversight for placement of a pulmonary artery catheter. In addition, repeated arterial and venous blood sampling is required for calculation of the arterial-venous difference in oxygen content. There are now less invasive methods to measure cardiac output, although new methods will have been validated by comparing the results with the classic Fick Method.

34.1.4   Required controls

A fundamental assumption is that pulmonary blood flow is equal to systemic blood flow, which will be true unless the subject has a cardiac or pulmonary shunt.

One of the primary assumptions of Fick's principle is that a steady state exists in the individual during testing. That is to say that proper implementation of this method requires, and should explicitly state in the methods description, that the conditions in the room include tightly controlled temperature, strict quiet by the investigator and subjects, and that the subject was in a maintained comfortable supine position for all measurements. Implicit in the achievement of steady state also assumes that there is a stable VO_2.

Achieving a stable VO_2 is not trivial, as many factors can alter it. The researcher should be mindful of recent meals, medications, exercise history, temperature, etc. Also, it is very important that the blood gas machines are properly and regularly calibrated.

34.1.5   Common problems or errors in literature and pitfalls in execution or interpretation

- Since the determination of cardiac output in this method is ultimately based on a calculation, it is vital that all factors in the calculations are carefully determined.

  - For example, some laboratories may choose to *estimate* whole-body oxygen consumption based on body surface area rather than *directly measuring* it; however, since the relationship of body surface area to oxygen consumption is highly variable this can introduce unnecessary error in calculations.

  - Care must to taken by the researcher to ensure that there are no air leaks around the mouthpiece, mask, or hood used to assess oxygen consumption.

- Since the technique is dependent upon accurate measurements of arterial and venous blood samples, care needs to be taken to obtain accurate samples, to properly collect the blood samples into tubes containing heparin and to store the blood in glass (not plastic) tubes prior to measurements.

- Researchers need to be mindful that carboxyhaemoglobin, as well as the low oxygen content in venous blood, may alter the accuracy of the readings (Grossman, 2006).

- Obtaining a good central venous sample, ensuring that this is actually taken from the correct catheter placement in the pulmonary artery.

- Repeated measurements should be made 3–4 times and averaged.

- Any heart valve defect or cardiac shunt that allows for blood backflow and/or mixing can negatively impact the method and make the data unreliable. Patients need to be screened for these dysfunctions.

34.2   Measurement of cardiac output by dilution (category A methods)

34.2.1   Thermo- and dye-dilution methods

The thermo- and dye-dilution methods to determine cardiac output are specific applications of Fick's principle. In this case, a bolus of non-toxic dye or cold fluid is injected into the vena cava or right atrium, where it mixes with blood in proportion to the blood flow in that area. The concentrations of the substance, or the temperature of the fluid at the site of injection, can then be compared to a distal site of collection, once blood is mixed with the indicator dye or cold fluid. This concept is similar to the Fick method, in which the content of oxygen in the pulmonary artery exiting the heart to the lungs was compared against the oxygen content in the systemic arterial circulation (Grossman, 2006).

The thermodilution method is currently commonly used and is considered by some to be the *de facto* standard method for assessment of cardiac output (Garcia *et al.*, 2011; Schroeder *et al.*, 2009). In this method, a cold bolus of water or saline is injected into the right atrium, where it mixes with blood and reduces the blood temperature. The temperature before and after injection of cold fluid in the right atrium is measured using a special catheter placed in the pulmonary artery; this catheter has a temperature-sensitive tip called a thermistor, to allow for measurements in real time of changes in blood temperature (Prabhu & Gulve, 2004).

When cardiac output is high, there is high blood flow through the right atrium, where the cold bolus is administered. The cold temperature bolus will be transported quickly to the distal site of measurement at the thermistor tip. Since there is minimal dilution of the cold injectate, there is little change in the temperature of the injected fluid. The researcher will measure a significant fall in temperature that is rapidly recovered. This is in contrast to the cases of low cardiac output, hence low blood flow, in which the cold temperature bolus will move slowly towards the measuring site and will become more diluted with that increased transit time. The blood-fluid mixture will take longer to reach the measurement site, and the recorded temperature drop will be less (Figure 34.1; Love *et al.*, 1990; Gawlinski, 2000).

The dye dilution method works by the same principles. The dye is injected into the pulmonary artery, mixes with blood in proportion to blood flow, then is sampled from a distal arterial site. The cardiac output is calculated from the dye concentration measured over time, and is inversely related to dye concentration (Prabhu & Gulve, 2004). This method is less commonly used than thermodilution.

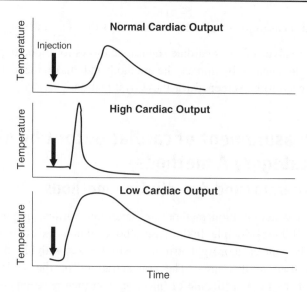

Figure 34.1 Example data from the thermodilution method of assessing cardiac output (modified after Love, Lough & Bloomquist, 1990). **Upper panel:** For normal cardiac output, a temperature curve obtained from the thermistor in the pulmonary artery would show a slight delay from cold-fluid injection until the change in temperature is detected. A smooth upstroke is expected, followed by a gradual return to baseline. **Middle panel:** For high cardiac output, there is a rapid detection of the temperature change and a small area under the curve, compared to normal cardiac output. **Lower panel:** For low cardiac output, there is a large area under the curve. Adapted from Love M, Lough ME, Bloomquist J. 1990. Cardiovascular laboratory assessment and diagnostic procedures. In Textbook of Critical Care Nursing: Dignosis and Management, ed. LA Thelan, JK Davie, and LD Urder, St - http://www.amazon.co.uk/Textbook-Critical-Care-Nursing-Management/dp/0801650038Louis, MO: CV Mosby.

34.2.2   Required controls

Some of the controls for this method are the same as those in the direct Fick method described above. This is especially true in the case that valve insufficiency and shunts will negatively impact data. In addition, it should also be assumed that the dye used meets the basic criteria, namely that the dye is non-toxic, mixes thoroughly with blood and is not altered in passage from injection to sampling. Commonly used dyes that meet these criteria are indocyanine green and lithium. Finally, it is absolutely essential that the dye or temperate injectates are properly mixed and are injected with precise repeatability.

34.2.3   Common problems or errors in literature and pitfalls in execution or interpretation

- Repeated measurements should be made 3–4 times and averaged.

- It is important to note that cardiac output and temperature of the blood are altered during phases of the respiratory cycle (Grossman, 2006). This is of particular concern in patients that are mechanically ventilated.

○ One strategy is that all repeated measurements used for averaging should be made at the same respiratory phase – typically, end-expiration. However, others use a different strategy to take multiple measurements at various phases of the respiratory cycle and average, thus giving an estimation of the overall range of cardiac output over the entire respiratory cycle (Schroeder *et al.*, 2009).

• Since this method is based on a dilution principle, the administration of the bolus of cold injectate or dye to be diluted is critical. Therefore, in the case of thermodilution, features of the cold fluid administration (e.g. type of fluid, volume, temperature, speed of bolus infusion) will all impact the data collected and must be controlled tightly and reported accurately. The duration over which the injectate is infusion should be done quickly and be highly repeatable.

• For thermodilution, it is typical that 5–10 ml of either room-temperature or ice-cold fluid is injected (Schroeder *et al.*, 2009). There is no difference in method accuracy with either temperature, so it is common that room-temperature fluid be used for ease of preparation compared to a controlled ice-cold temperature.

• Any heart valve defect that allows for blood backflow can negatively impact the method and make the data unreliable. Patients need to be screened for valve dysfunction.

• The investigator needs to be very familiar with what a 'good' dilution curve looks like under a variety of conditions (see Figure 34.1), so that one can troubleshoot in real time and collect additional data as necessary during the same study if 'bad curves' are generated. Average data from multiple 'good curves' is an important assumption made when reviewing data.

34.3 Non-invasive cardiac output (category B methods)

Given the invasive nature, cost and advanced technical expertise needed to perform invasive cardiac output measurements, there has been great interest in recent years to develop and validate non-invasive methods. Several of these are briefly discussed below, but this is not an exhaustive list (Schroeder *et al.*, 2009).

34.3.1 Partial carbon dioxide gas re-breathing

In yet another variant of the Fick principle, carbon dioxide diffusing through the lungs is assessed non-invasively to determine cardiac output. This is a desirable alternative to the invasive methods and does not require expensive equipment like ultrasound. Ultimately, this method is assessing pulmonary blood flow as a means to determine cardiac output (Schroeder *et al.*, 2009). The benefit of this approach is that the need for invasive catheters to measure venous and arterial blood directly is replaced by the measurement of CO_2 from the breath during intermittent partial CO_2 re-breathing.

Briefly, the measurement system includes a CO_2 sensor, an airflow sensor and a pulse oximeter. CO_2 production and end-tidal CO_2 are measured at rest and during repeated bouts of partial re-breathing. Partial re-breathing is achieved by a valve in the system used to increase dead space transiently, resulting in reduced CO_2 elimination from the lungs and a temporary increase in end tidal CO_2. The increase in CO_2 during re-breathing is used to determine cardiac output (Berton & Cholley, 2002). The venous CO_2 can be determined by comparing the expired and inspired gases for CO_2 content, rather than through invasive blood measurements as in other methods. The arterial CO_2 content is estimated by multiplying the slope of the CO_2 dissociation curve and the change in end tidal CO_2 (Young & Low, 2010).

Required controls   Since the main measurements used to determine cardiac output are derived from breath analysis, it is essential that the system is free from leaks. The subjects should be resting in a comfortable position to ensure that steady state measurements are taken. Changes in metabolism and muscle work can alter CO_2, so the trials should be done with the subject fasting, without having performed heavy exercise for 24 hours.

Common problems or errors in literature and pitfalls in execution or interpretation

- Special consideration is required for patients with pulmonary disease and severe lung injury (Young & Low, 2010).

- It is essential that the analyzer for end tidal CO_2 is properly calibrated.

- The researcher needs to be mindful of anything that interferes with the end tidal CO_2. It would be assumed by the reader that there was no yawn, sigh, cough, etc. during data collection that would interfere with obtaining clean data.

- The equipment must be assumed to have been properly calibrated, any masks used need to be tested for a good seal with no leaks, and any bag used for gas breathing must have been checked each use for no holes or leaks.

- Depending on the technique, the investigator must know what 'good data' looks like in real time.

34.3.2   Ultrasound-based methods of cardiac output

Cardiac ultrasound is used to view the heart chambers directly and to measure the blood volumes contained in the chambers in real time. Cardiac output can be measured using this method, combined with pulse wave Doppler applied supra-sternally to measure blood flow velocity through the mitral or aortic valve or the left ventricular outflow tract (Garcia *et al.*, 2011). There are also minimally invasive usage of Doppler which includes pulmonary artery, transtracheal and transoesophageal approaches (Berton & Cholley, 2002; Garcia *et al.*, 2011; Schroeder *et al.*, 2009). The blood velocity measurement is related to the Doppler principle.

Briefly, ultrasound waves emitted from the measurement device are reflected back to the emitter at a different frequency after they encounter a moving object, such as red blood cells in the artery. The shift in frequency of the reflected ultrasound waves is related to the velocity of the encountered moving object, thus allowing for a calculation of blood flow velocity based on analysis of the reflected waves. This calculated blood velocity is then used in a further calculation, by taking the integral of the velocity over time to determine stroke volume which, when combined with heart rate, will provide cardiac output. This is an accurate method to assess cardiac function.

Required controls This method requires expensive equipment and advanced training, both to perform the measurements and to analyze the data. The correct selection of ultrasound probe and appropriate settings must be assumed. The angle position of the probe that emits is important, as the angle of waves hitting the red blood cells also impacts the frequency shift observed (Schroeder *et al.*, 2009).

Common problems or errors in literature and pitfalls in execution or interpretation There can be problems that arise in obtaining a good image, due to variations in anatomy and obesity. Since the data is based on a good resolution with clearly discernable wall edges and valves, it is essential that image collection is done with great care.

34.3.3 Calculations of cardiac output from arterial pressure waveforms

The simplest non-invasive method to assess cardiac output is to analyze the arterial pulse waveform (Broch *et al.*, 2012). The arterial pulse waveform can be obtained using non-invasive measurement, such as the finger blood pressure cuff known as the Finometer® or NexFin®, or a method known as arterial tonometry, where a sensitive probe is placed against the skin above the pulsatile artery. Arterial waveforms can also be obtained invasively, using intra-arterial catheters. A computer program uses data modelling to determine cardiac output from certain features of the pressure waveform, namely during the systolic part of the curve that is bordered by the end of diastole and the end of the systolic ejection phase (Prabhu & Gulve, 2004). This technique has been validated by comparison with both the classic Fick method and thermodilution.

Required controls

- This method gives an estimation of cardiac output and relies upon the appropriate values for subject height and age to be entered into the software algorithm that determines the values for cardiac output. It is essential, then, that accurate values are obtained for each subject, and that the correct values are typed into the system.

- These non-invasive devices have a height correction feature to correct for the fact that the pressures are taken from the finger, not at heart level. It is essential that the height correction feature is used.

- Since the data are derived from the arterial pressure waveform, it is critical that the data are collected with strict calibration, correct zeroing and adequate harmonic features (Garcia *et al.*, 2011).

Common problems or errors in literature and pitfalls in execution or interpretation:

- This method is not suitable for critically ill patients.

- The data collected using this method are most accurate in determining changes in cardiac output from control during an interventions. The method is less accurate for determining absolute values for baseline pressures or for comparing baseline pressures across subjects.

- There are many reasons why the arterial wave can appear skewed. The investigator must have ensured that an accurate waveform is collected and analyzed. For example, excessive vasoconstriction can result in an exaggerated reflected peripheral wave form that would impact calculations of cardiac output.

34.4   Complementary and/or adjunct techniques:

Primer 35: Blood flow measurement by venous occlusion plethysmography.
Lower body negative pressure.
Spirometry.
Blood pressure measures.
ECG.
Blood gas analysis.

Further reading and resources

Berton, C. & Cholley, B. (2002). Equipment review: new techniques for cardiac output measurement – oesophageal Doppler, Fick principle using carbon dioxide, and pulse contour analysis. *Critical Care* **6**, 216–221.

Broch, O., Renner, J., Gruenewald, M., Meybohm, P., Schottler, J., Caliebe, A., Steinfath, M., Malbrain, M. & Bein, B. (2012). A comparison of the Nexfin® and transcardiopulmonary thermodilution to estimate cardiac output during coronary artery surgery. *Anaesthesia* **67**, 377–383.

Garcia, X., Mateu, L., Maynar, J., Mercadal, J., Ochagavia, A. & Ferrandiz, A. (2011). Estimating cardiac output. Utility in the clinical practice. Available invasive and non-invasive monitoring. *Medicina Intensiva* **35**, 552–61.

Gawlinski, A. (2000). Measuring cardiac output: intermittent bolus thermodilution method. *Critical Care Nurse* **20**, 118–120, 122–124.

Grossman, W. (2006). Blood flow measurement: cardiac output and vascular resistance. In: Balm, D.S. & Grossman, W. (Eds.). *Grossman's Cardiac Catheterization, Angiography and Intervention*. Williams and Wilkins, Baltimore, MA.

Love, M., Lough, M.E. & Bloomquist, J. (1990). Cardiovascular laboratory assessment and diagnostic procedures. In: Thelan, L.A., Davie, J.K. & Urder, L.D. (Eds.), *Textbook of Critical Care Nursing: Diagnosis and Management*. CV Mosby, St Louis, MO.

Pearl, R.G., Rosenthal, M.H., Nielson, L., Ashton, J.P. & Brown, B.W.Jr. (1986). Effect of injectate volume and temperature on thermodilution cardiac output determination. *Anesthesiology* **64**, 798–801.

Prabhu, M. & Gulve, A. (2004). Cardiac Output Measurement. In: *Anaesthesia and intensive care medicine*, pp. 49–52. The Medicine Publishing Company Ltd.

Schroeder, R.A., Barbeito, A., Bar-Yosef, S. & Mark, J.B. (2009). Cardiovascular Monitoring. In: Miller, R.D. (Ed.), *Miller's Anesthesia*, pp. 1267–1328. Churchill Livingstone.

Young, B.P. & Low, L.L. (2010). Noninvasive monitoring cardiac output using partial CO_2 rebreathing. *Critical Care Clinics* **26**, 383–392.

35
Measuring Peripheral Blood Flow in Humans

Erica A Wehrwein[1] and Michael J Joyner[2]

[1]*Physiology, Mayo Clinic, Minnesota, USA*
[2]*Anesthesia Research, Mayo Clinic, Minnesota, USA*

35.1 Basic 'how-to-do' and 'why-do' section

Blood flow is the quantity of blood that passes by a given point in the circulation in a particular time; hence, it is reported as volume per unit time – commonly ml/min. The blood flow, when considered over the whole circulation, is roughly equivalent to cardiac output. Thus, if cardiac output is determined to be 5 litres/min, then total blood flow is 5,000 ml/min.

There are several non-invasive methods to assess aspects of peripheral blood flow. The assessment of peripheral blood flow in the limbs can offer insight into total peripheral resistance, blood pressure, vasodilation capacity and the effects of numerous interventions, including exercise. While assessment of peripheral blood flow can be done at rest, it is often also performed during an intervention such as exercise, drug infusion, psychological stress, and many others. Regardless of the intervention, the basic methodological considerations below are valid for accurate and repeatable data collection. This primer will cover two approaches:

- Venous occlusion plethysmography
- Doppler ultrasound

35.2 Venous occlusion plethysmography

Venous occlusion plethysmography (VOP) is a commonly used non-invasive method used to assess limb blood flow that has been well validated against other

Essential Guide to Reading Biomedical Papers: Recognising and Interpreting Best Practice, First Edition.
Edited by Phil Langton.
© 2013 by John Wiley & Sons, Ltd. Published 2013 by John Wiley & Sons, Ltd.

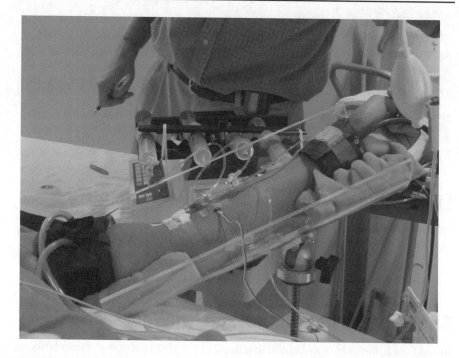

Figure 35.1 Photo of venous occlusion plethysmography set-up. This photo shows the upper and lower arm cuffs, with the mercury strain gauge around the forearm just below the elbow. Note that the hand is elevated to promote venous drainage. Courtesy of Professor M. J. Joyner.

methods under a range of conditions (Wikinson & Webb, 2001). It is often used to measure blood flow in the arm, but can also be used in the leg. This primer will focus on the arm measurements. VOP has a simple set up using a blood pressure cuff on the upper arm that is cyclically inflated to a pressure that is above venous but below diastolic arterial pressure (Figure 35.1).

During a period of measurement, the pressure in the cuff will be cycled, with each cycle consisting of:

- Inflation 7–10 seconds

- Deflation 5–7 seconds

The cuff, which is applied to the upper arm, is rapidly inflated to between 40–60 mmHg, enough to [1]*occlude* venous drainage from the limb, hence the name 'venous occlusion'.

During the period of inflation, arterial blood continues to enter the arm, as the occluding pressure in the cuff is far below the lowest (diastolic) arterial blood pressure. Moreover, this blood pools within the veins of the limb, trapped by the

---

[1] **Occlude** – meaning to shut or block.

inflated cuff on the upper arm cuff. This accumulation of blood, being a liquid and non-compressible, effectively increases the volume of the arm below the cuff – essentially, the arm 'swells up'. The increase in blood (and arm) volume is measured as an increase in forearm circumference, using a transducer called a 'strain gauge' (typically mercury in silastic tubing), which is wrapped around the widest part of the forearm.

The typical mercury strain gauge response to being stretched is to increase its electrical resistance. Typically, this change is converted into a voltage output signal (a constant current will result in a voltage signal that varies in direct proportion to the resistance of the strain gauge), and thus digitized and recorded.

After not more that ten seconds of inflation, the cuff is rapidly deflated for 5–7 seconds, allowing the accumulated venous blood to pass the cuff and restore the normal balance of arterial and venous blood flow. This is, in part, facilitated by having the subject's arm slightly elevated. During the period of deflation, the volume of the arm returns to normal. This cyclic inflation/deflation is applied for several minutes during data acquisition.

It should be obvious that the rate at which blood accumulates in the limb during venous occlusion will also reflect the rate at which forearm volume changes. If blood flow increases (remember, flow = volume per unit time), the rate at which the forearm volume increases during occlusion will, itself, also increase. Flows determined using this method are reported as ml per 100 ml of forearm volume per minute if a calibration has been performed – otherwise only relative changes in strain output will be reported (Figure 35.2).

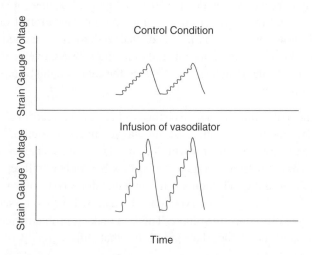

Figure 35.2 Example data from venous occlusion plethysmography. Upper panel: a control subject with baseline blood flow measured while supine. Lower panel: a subject following infusion of a vasodilator to increase blood flow. The increase in flow is observed as an increase in slope compared to control conditions. Courtesy of Professor M. J. Joyner.

35.2.1   Required controls

Since the hand has a complex circulation and the hand vessels tend to have a pharmacology that differs from vessels in the forearm, it is typical that the hand is not included in this measurement. The hand circulation is excluded by the use of a wrist cuff inflated to a pressure above systolic (super-systolic) pressure, and this cuff should be rapidly inflated 60 seconds prior to the measurement to ensure that forearm flow is stabilized. The hand should be elevated on a pillow or towels to keep it above heart level and to ensure that venous drainage from the forearm is adequate once the upper arm cuff is deflated (Wikinson & Webb, 2001). The subject should be instructed to remain very still and not contract the limb during measurements.

It is important to use a device to ensure that the cuffs are automatically rapidly inflated and deflated rather than the slower manual approach. The study should be conducted in a quiet, temperature-controlled room with the subject recumbent or supine.

35.2.2   Common problems or errors in literature and pitfalls in execution or interpretation

- This type of blood flow data is often analyzed by a researcher manually. Therefore, it is important that for any given study, all blood flow data obtained with VOP is analyzed by the same individual to prevent bias.

- As shown in the cartoon of idealized data in Figure 35.2, there are multiple flow peaks collected for each period. Pending any obvious artefact or error in the peak, all flows will be marked and analyzed to be averaged together. A common source of artefact is cuff inflation and subtle movement of the strain gauge output cable. In the case of the later, the cable should be taped to prevent excessive movement.

- Reporting of absolute or relative changes: changes in forearm volume, as noted above, are monitored with a strain gauge that will output a voltage in proportion to increase in forearm volume. There are instances where investigators will not assign a value to blood flow but, rather, will report this strain gauge voltage data and compare the voltage slope (rate of volume change) in control and during an intervention. The data will be expressed in terms of a percentage difference in voltage. This can make it problematical to compare between subjects, if they have forearms that differ greatly in volume (i.e. muscle mass).

- Differences in calibration: data collected from a calibrated system will allow the experimenter to express the changes in blood flow in terms of ml per

100 ml of forearm volume per minute. Forearm volume can be determined using:

○ water displacement: the same principle as Archimedes;

○ estimation: assuming the forearm to be a truncated cone.

Gauges can be manually calibrated using 'holders' adjusted to cause a known change in gauge length (Whitney, 1953). Electronic calibration is now a standard feature of modern plethysmographs (Hokanson).

• If data is collected on forearm drug infusion, it must be noted that a forearm volume measurement was obtained for each individual being studied, in order to ensure equivalent dosing.

• If the strain gauge is not properly sized to the individual being studied, this can lead to inaccurate results.

• Arm phenotype is highly variable and may cause difficulties in recordings. For example, if the arm is highly tapered, it can result in slippage of the strain gauge, so often the strain gauge is taped to the forearm.

• The pressurized cuffs are assumed to have been checked for leaks to ensure that proper pressure in maintained during the measurements.

• It is possible that electrical short-circuits can develop in the strain gauge, leading to poor data quality.

• It is assumed that the investigator has controlled for all factors that can result in a non-linear slope of the curve (linear slopes are shown in Figure 35.2). The features that are most commonly controlled are factors like cuff inflation artefact and the subject remaining stationary during recordings.

35.3   Ultrasound-based methods of blood flow

Peripheral limb blood flow can also be assessed using ultrasound which offers the advantage of the determination of the diameter of the large conduit vessels as well as assessing blood flow velocity with Doppler as described above (Figure 35.1; Thrush & Hartshorne, 2005). The basic premise is that estimates of mean blood velocity (determined by Doppler) are combined with the vessel cross-sectional area measurements (using B-mode ultrasound) to reveal limb blood flow (volume/time). This method can be applied non-invasively and is commonly done over the brachial, femoral and popliteal arteries. It is beyond the scope of this primer to cover in detail the theory of ultrasound and the appropriate selection of probe and settings; there are many references that cover these topics thoroughly (Thrush & Hartshorne, 2005). This method requires expensive equipment and technical expertise.

35.3.1   Required controls

- The data obtained from this approach is dependent upon a variety of conditions that should be applied consistently to all subjects once optimized. This includes ultrasound beam width and angle of probe. Probe angle is of particular importance.

- The use of a high-resolution probe is essential for obtaining images with adequate resolution, in order to get accurate arterial wall images where the vessels walls can be clearly discerned. This is vital, since the vessel cross-sectional area is used (along with velocity) in calculations to determine flow.

- Since vessel diameter changes slightly as the beating heart expels a volume of blood (cardiac cycle), in some cases it is necessary to 'gate' the data acquisition to the cardiac cycle. This can be done by capturing images of vessels diameter during a particular phase of the cardiac cycle (e.g. end diastole) (Harris *et al.*, 2010).

- When making measurements to compare baseline to an experimental condition such as exercise or drug infusion, it is important that the above parameters are strictly maintained and that the probe angle and position are consistent between pre- and post-measurements.

- It is preferred that a single investigator performs the data analysis, in order to ensure that consistent measurements of vessel diameter are made and any bias is avoided. This is essential, since the vessel diameter is used to calculate flow and any differences in manual demarcation of the vessel wall will influence the calculation.

- Since many medications act directly or indirectly on the cardiovascular system and have the ability to alter the measurements of velocity and the diameter of the vessel walls, it is important to study individuals who are not taking these medications, or who have stopped taking the medications in advance of the study to allow for a wash-out period. When it is not possible for subjects to be off their medications, this should be disclosed as a potentially con-founding variable.

35.3.2   Common problems or errors in literature and pitfalls in execution or interpretation

- Investigators need to report raw blood flow values, blood pressure and heart rate, so that the reader has all the necessary information to calculate con-ductance and resistance, as opposed to the paper reporting only the calculated values. These values are sometimes not reported.

- Vascular resistance offers a measure of impedance to flow and is a *calculated* variable. When reporting data on resistance, it is necessary also to include the *collected* data on conductance or flow that was used to determine the resistance value. Similarly, when using ultrasound, the investigators should report the velocity and vessel area that was used to determine flow.

- The most accurate determination of vessel diameter is to use an ultrasound beam width that is larger than the vessel diameter. However, this condition is often not met when studying the larger conduit vessels commonly assessed. In this case, there will be an overestimation of velocity that will be carried though any calculations based on velocity, such as flow.

- When comparing the relative difference in velocity and flow across experimental conditions (e.g. drug infusion) between two groups, one must consider whether or not baseline values are the same between groups. If only the differences in the experimental values are considered between subjects, the investigator may not have a clear understanding of the effect of the intervention. For example, a study reports that there was no difference in peak blood flow during a vasodilator drug infusion between group A and group B, yet the starting baseline values for group A were significantly lower that group B. That means that the overall change from baseline to peak in group A was actually much larger than that which occurred in group B, so therefore there was a larger effect of the drug in group A. This finding would be lost if only the experimental values (and not the control) were reported.

35.4　Complementary and/or adjunct techniques

Primer 34: Cardiac output
ECG
Blood pressure
Oxygen consumption

Further reading and resources

Harris, R.A., Nishiyama, S.K., Wray, D.W. & Richardson, R.S. (2010). Ultrasound assessment of flow-mediated dilation. *Hypertension* **55**, 1075–1085.

Hokanson, D.E., Sumner, D.S. & Strandness, D.E. (2000). An electrically calibrated plethysmograph for direct measurement of limb blood flow. *IEEE Transactions on Biomedical Engineering* **22**, 21–25.

Hokanson EC5R strain gauge plethysmograph technical manual: www.deh-inc.com/documents/Forearm%20Bloodflow.pdf

Joyner, M.J., Dietz, N.M. & Shepherd, J.T. (2001). From Belfast to Mayo and beyond: the use and future of plethysmography to study blood flow in human limbs. *Journal of Applied Physiology* **91**, 2431–2441.

Thrush, A. & Hartshorne, T. (2005). Blood flow and its appearance on color flow imaging. In: Thrush, A. & Hartshorne, T. (Eds.), *Peripheral Vascular Ultrasound: How, Why, and When*, pp. 49–62. Churchill Livingstone, New York.

Wilkinson, I.B. & Webb, D.J. (2001). Venous occlusion plethysmography in cardiovascular research: methodology and clinical applications. *British Journal of Clinical Pharmacology* **52**, 631–646.

Whitney, R.J. (1953). The measurement of volume changes in human limbs. *Journal of Physiology* **121**, 1–27.

Index

Note: Figures and Tables are indicated by *italic page numbers*, footnotes by suffix '[n]'

Essential Guide to Reading Biomedical Papers: Recognising and Interpreting Best Practice, First Edition.
Edited by Phil Langton.
© 2013 by John Wiley & Sons, Ltd. Published 2013 by John Wiley & Sons, Ltd.